ALTERED STATES

A Field Guide to Your Brain on Drugs

By Jeremy Wolfe

Wolfe Bentley Publishing
Boulder, Colorado

This book is for educational and harm reduction purposes only. It does not constitute medical, legal, or mental health advice, and nothing in these pages should be interpreted as encouraging or endorsing illegal activity. Psychoactive substances affect individuals differently based on biology, dosage, set and setting, and pre-existing conditions. Readers with health concerns, particularly related to mental illness or medication interactions, should consult a licensed professional. Possession and use of many substances discussed here may be illegal in your jurisdiction. Know your local laws. The author and publisher accept no responsibility for choices made by readers.

AI tools were used for research verification, editorial support, and image development. The author is solely responsible for the writing, ideas, and content of this book.

Library of Congress Control Number: 2026911526
ISBN 979-8-218-93672-3 (Paperback)
ISBN 979-8-9959700-0-2 (Hardcover)

10 9 8 7 6 5 4 3 2 1

Wolfe Bentley Publishing – Boulder, Colorado, USA
Printed in the United States of America

Dedication

This book is dedicated to everyone who held up the heavy posterboard slides at my Burning Man lectures.

Next time, I'll buy an easel.

Table of Contents

Disclaimer

This book is for educational purposes only. It is not intended to encourage or endorse the use of any illegal substances. Instead, it strives to provide honest, accurate, and compassionate information about how people can reduce harm in settings where use may already occur.

Psychedelics, empathogens, dissociatives, and other psychoactive substances can affect individuals differently based on biology, psychology, dosage, set and setting, and pre-existing medical conditions. While some people report deep healing and insight, others experience confusion, fear, or long-term psychological effects.

Nothing in this book should be interpreted as medical or legal advice. The information presented does not substitute for consultation with a qualified legal, medical, or mental health professional. If you have any health concerns, especially related to mental illness, medication interactions, or substance use, **speak with a licensed provider.**

The information contained in this book represents the author's best understanding of community practices at the time of writing. New substances come out every day, and every day science brings us new understanding of our brains. Keep up to date, do your own research, and educate yourself. Your body is your own.

In many jurisdictions, possessing or using the substances discussed here is illegal. Know the laws in your region. The author and publishers accept no responsibility for the choices readers make.

Finally, harm reduction is not risk elimination. Even with best practices, there is no such thing as a completely safe drug. The goal is to help people stay informed, take these substances seriously, and make conscious, intentional decisions that minimize risk to themselves and others.

Special Thanks

This book wouldn't be possible without the sponsorship and support of my Kickstarter backers. Thank you for believing! I'm sincerely grateful to each of you. This book is yours as much as it is mine. Special thanks to the following:

Sharon Alexander
Maureen Wolfe
Howard Wolfe
Mike Wolfe
Christy Wolfe
Surprise Paradiso
Arthur Bentley
Rich Matney
Brad Chester
Steve Sommer
Dr. Matthew Luckett
Shandi Lenaburg
Diane Klein
Kyle "Squirrel" Herd
Eric Beall
Amanda McKim
Margot Ruthardt
Ben Alexandra
Christine Topinka
Robin Tolentino
The LeBlanc Family
Donghyeok Yang
Tyler Dane
adi
Livi E Wilde
Oz Augé
Sean Connell
Keith Wilcox

Preface

There's a paradox in the world of drug education. Many people who use substances recreationally or spiritually know far more about the science behind them than those who write the laws prohibiting them. And yet, much of this knowledge is scattered, hard to verify, or passed down like urban legends.

That's why this book needed to be written.

Mind-expanding drugs such as psychedelics, stimulants, dissociatives, and empathogens interact directly with the systems that shape your thoughts, your emotions, and your sense of self. If you're going to engage with substances that alter your brain chemistry, it's only reasonable (and maybe even moral) to understand what they're doing, how they do it, and how to keep yourself and others safe in the process.

That's harm reduction, and it starts from a simple premise: people will make their own choices, so let's make sure they don't die from them. It's the seatbelt of drug culture. Education, compassion, and realism, applied where they're needed most.

You may notice this book doesn't attempt to cover every substance that exists. Instead, I chose a set that reflects the compounds people encounter most often, the ones with enough research to explain what they actually do, and the ones where better information genuinely changes outcomes.

In the pages that follow, you'll find a blend of biochemistry, cultural history, supplementation science, safety guidelines, and street-level wisdom. We'll talk about how neuro-transmitters work, why your jaw clenches on MDMA, how LSD rewires your thalamic filters, and why you should never, ever mix LSD with lithium unless you're trying to meet the local paramedics.

There will be science. There will be slang. There will be caution, humor, and a lot of swallowing large vitamins.

This book doesn't claim to make drug use "*safe*." Let me be perfectly clear: **drugs are dangerous**. So is driving. So is swimming. So is falling in love. Danger is a part of life, but reckless ignorance doesn't have to be. Nothing truly worth doing in life is without risk, but this book can help you be smart about it.

If you're going to explore altered states of consciousness, whether once in a lifetime or on a monthly cycle, this guide is for you.

Listen to your body, and respect the chemistry.

Let's begin.

- Jeremy "*Dr. Awkward*" Wolfe[1]

[1] Not an actual doctor, but someone who reads the footnotes.

Before We Start...

"It's no use going back to yesterday, because
I was a different person then."

This book is about harm reduction.

It is a guide built on the assumption that people will make their own choices, and that those choices deserve to be informed by real science, honest conversation, and genuine care. It is not a dare, a challenge, or a "how-to-do-drugs" manual.

Before we talk about how to use substances more safely, we need to talk about something that lives on the other side of the conversation. Something that doesn't always announce itself, and doesn't always look the way you expect.

We need to talk about **addiction**.

Addiction is a neurological condition with roots in brain chemistry, genetics, trauma, environment, and the very reward systems we will spend this entire book exploring. It is not a moral failure, a weakness, or a personality flaw. The same dopamine pathways that make music sound incredible on MDMA are the ones that can, under certain conditions, trap someone in a cycle they didn't choose and can't easily escape.

Not everyone who uses substances develops a problem. Most people don't. But for some, what starts as curiosity, recreation, or even healing can shift into something harder to control.

That shift is often invisible at first. You might notice that you're thinking about the next session before the last one has fully settled. You might find that you're using more often than you planned, or that the reasons have quietly changed from celebration to escape. You might realize that the substance isn't adding to your life anymore. It's filling a hole.

If any of that sounds familiar, this is where the book reminds you that you deserve support, not shame. There are people and organizations built specifically to help, in the United States, Canada, Europe, and beyond. They will not judge you. Many of them have walked the same road.

Harm reduction works on a spectrum. At one end is making a dangerous thing slightly less dangerous. At the other end is recognizing when the thing itself has become the problem, and finding the courage to ask for help. Both are valid. Both take strength. Both deserve respect.

If you picked up this book because you're curious, welcome. If you picked it up because you're cautious, good. If you picked it up because you're struggling, please know that help exists for exactly this moment, and there is no version of your story that is beyond repair. Turn to the appendices at the back of this book for a full list of crisis lines, treatment resources, and support organizations sorted by region.

That's the whole point of this book: **you** are worth protecting.

Who Are You? – A Biochemical Perspective

"Who in the world am I? Ah, that's the great puzzle."

Who *are* you? It's a difficult question to answer, at the heart of it. It's easy to say your name, but it's hard to say what makes any of us *us.*

This conversation isn't new. People have been asking this same question for thousands of years. Philosophers, priests, and poets have taken their turns at the question, pointing variously to the soul, the body, the mind, or the memories that hold us together.

Neuroscience offers a different model, one that is both humbling and empowering. From the perspective of modern brain science, the "self" is what emerges when several biological systems run at once, like the ones that handle your senses and emotions, your memory, and the running map of your own body. The whole thing feeds back into itself constantly, creating the experience of being *you*. These many sensory systems may often add up to one continuous experience, but that unity is more fragile than it feels, and it can be shifted.

This sits at the center of understanding how substances affect you and why harm reduction matters. None of what you think or feel or do is floating in some intangible spiritual cloud. It's physical. It's chemical. It's electrical. You are an ensemble of signals and frequencies, neurotransmitters and hormones, voltage shifts and membrane potentials. Every memory, every emotion, every flicker of insight is rooted in your brain's internal wiring and your body's ongoing chemical conversation.

What Is Identity?

Is your identity your name? Your memories? Your personality? We tend to think of the "self" as something stable and core. The truth is that who you are, how you feel, think, and interpret the world, is all built from biochemistry. Your sense of "me" is an ongoing process emerging from patterns of neurotransmitters and electrical signaling. We are ever-changing systems of activity, dynamic and biological.

Everything you feel, every sight, sound, or stray thought is constructed by your brain out of raw sensory data. That perception is shaped by multiple factors. Levels of key neurotransmitters (like serotonin, dopamine, and GABA), the coordination of brain networks (like the default mode network), and the library of memories and associations your brain has built over a lifetime all come together to mold your view of the world. Your brain filters and reconstructs this data moment by moment to build what you believe to be real. When you take a psychoactive substance, you're not necessarily "seeing something fake." You're running a different mode of processing.

Psychoactive drugs change your perception by changing the chemistry underneath. A drug might bind to a receptor, or block one. It might change how much of a neurotransmitter is released, or how fast the leftovers get mopped up. Others reach further in, changing how distant brain regions talk to each other or retiming your brainwaves entirely. MDMA dumps serotonin, dopamine, and norepinephrine into your system all at once, which lifts your mood and makes you feel closer to the people around you. LSD lands on 5-HT2A receptors and switches off your brain's usual filters, so sensory and emotional input comes pouring in. Ketamine blocks NMDA receptors and pries your perception loose from your emotions, sometimes all the way to full dissociation.

What you get is a temporary shift in how you process information, and so a shift in who you are. Under the influence, you might feel more loving, or fearful, or cosmic and ancient, or strangely childlike. None of those are "less real." They're just less familiar.

You don't have a single self. You're a constellation of tendencies and moods and patterns, not a static thing. Drugs don't implant something new. They turn the volume up or down on what's already there. That's why a psychedelic can lead one person toward a spiritual opening and another toward deep anxiety, or why a stimulant sharpens focus for you and tips your friend into mania. Your inner symphony plays differently depending on which instruments get pushed forward.

Understanding this is important for how you see yourself, and how you see the substances you put in your body.

The Brain Signal Loop

Every moment, your brain takes in a flood of sensory data: light, sound, the feel of your clothes, the temperature of the room. It filters and sorts those raw signals before you ever become aware of them. It loops each one through, checking what it's getting against what it expected and quietly throwing out the noise. That processing is what gives you a coherent world instead of chaos. It's also what gives you the ability to reflect on what you're perceiving and choose what to do about it.

Step 1: Sensory Stimulus

The process begins with something happening around you, like light striking your retina, sound waves entering your ear, or pressure on your skin. These physical signals are converted into electrical impulses by sensory neurons and sent to the brain for processing.

Step 2: Selective Filtering

Most of this raw data first passes through the thalamus, a relay station that routes the signal to the appropriate sensory cortex (e.g., visual, auditory, tactile). The thalamus acts as a filter and regulates the flow, enhancing certain signals and suppressing others based on what's relevant in the moment.

Try this: what do your toes feel like right now? You probably had no awareness of them until you read that, even though they've been sending sensory data the whole time. That's thanks to your thalamus.

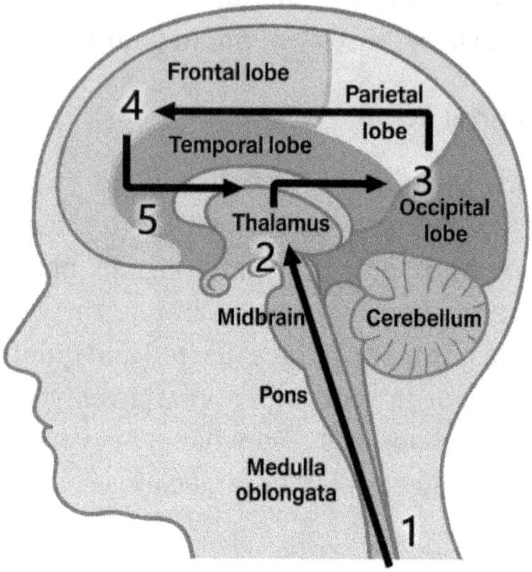

Feedback Loop

Before a signal reaches conscious awareness, it passes another early checkpoint, the amygdala, which sizes up its emotional weight. Is this safe, or is it a threat? The amygdala can trigger a fast physical response, a jump in heart rate or a dump of stress hormones, before you're even aware of what you're reacting to. That pre-conscious filtering helps you dodge danger quickly, but it also quietly colors how you read everything else.

Step 3: Signal Processing

The signal then gets routed to its sensory center to be turned into something the brain can interpret. Visual signals go to the occipital lobe at the back of your brain, where a picture forms. Auditory signals travel to the temporal lobe and become sound.

From there it moves into deeper integration pathways in the midbrain, where information from different senses gets combined. This is where the brain starts building a model of what's happening and how it connects to what came before.

The hippocampus, closely connected to the amygdala, plays a central role in linking the input to memory. It answers the question: Have I done something like this before? This process gives context and meaning to the signal and influences what you expect to happen next.

Step 4: Conscious Interpretation

Next, the signal reaches the frontal cortex, the seat of reasoning and self-reflection. This is where you consciously interpret the experience. You might ask, "What is this, and how should I respond?" The frontal cortex weighs risks against rewards and decides whether to act or just watch.

This stage is also where intention, morality, and self-awareness emerge. While earlier stages of the loop may generate reflexes or emotions, it is the frontal cortex that gives you the power to override them. It's what lets you talk instead of shout, pause instead of react.

Step 5: Feedback and Refinement

After all that, the brain usually goes looking for more. It sends attention back to the senses ("look again," "listen harder") or pings the memory circuits ("when have I felt this before?"). The loop keeps adjusting. Experience gets built moment by moment and updated in real time as new information is integrated.

The whole cycle runs in milliseconds, and every drug in this book leans on it somewhere. This processing loop is where sensation turns into experience and where chemistry turns into meaning. Understanding it gives you power, not just over how you use drugs but over how you handle your own mind.

Every thought you have about who you are runs through that same cycle. Input, then feeling, then memory, then meaning. Change the loop and you change the self. That's the reason this book exists. If you're going to mess with the machinery of perception, you ought to know how it runs.

Oh Snap, the Synapse

You don't have to be a neuroscientist to follow how drugs affect you, but you do need to understand the basics. Once you know how your brain passes messages, and what's carrying them, it gets a lot clearer how nudging that chemistry can ripple out into your mood and even the way time feels.

Get the mechanics of one neuron handing off to the next, and you'll see why a little MDMA can feel like the best hug of your life, and why the wrong mix of substances can make the whole thing short-circuit.

At the center of your nervous system are *neurons*, the specialized cells that send and receive information. They line up end-to-end without ever quite touching. Instead, they communicate by releasing chemicals in the tiny gaps between them. These gaps are called *synapses*, and are where most of the chemistry in this book occurs.

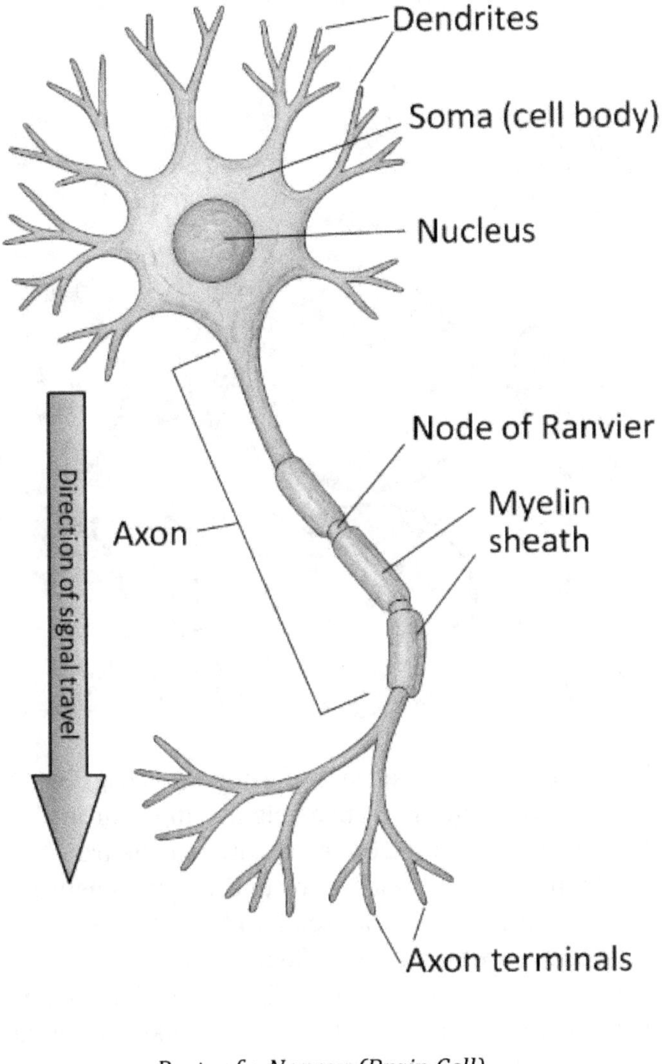

Parts of a Neuron (Brain Cell)

How Brain Cells Communicate

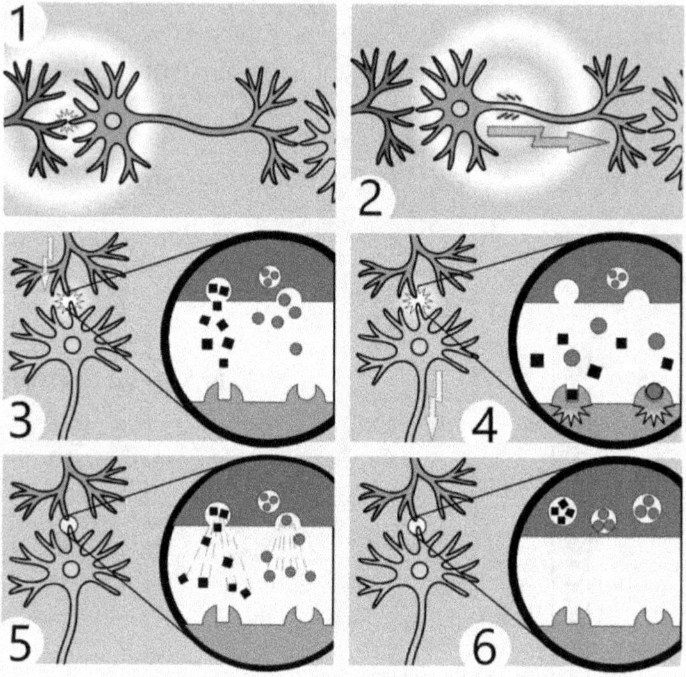

1) A receptor on the surface of a neuron is triggered by neurotransmitters from a neighboring neuron.
2) An electrical impulse travels down the neuron's axon tail.
3) When the signal reaches the end, it causes neurotransmitters to be released into the synapse.
4) These neurotransmitters float across the gap and bind to receptors on the next neuron, passing along the message.
5) Once the message is delivered, reuptake pumps suck the chemicals back in for recycling.
6) Any stray chemicals are cleared; the synapse resets. The cell is ready for the next signal.

The Only Things You Like

If your brain is a conversation, neurotransmitters are the words. They're the chemicals that carry signals from one brain cell to the next, and they shape everything from your hunger to your sense of self. Every thought you've ever had started as one of these messages crossing a gap. Let's meet the three we'll talk about the most: *dopamine, serotonin,* and *norepinephrine.*

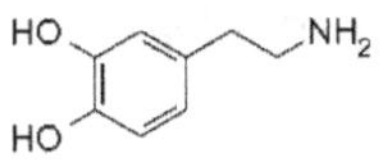

Dopamine

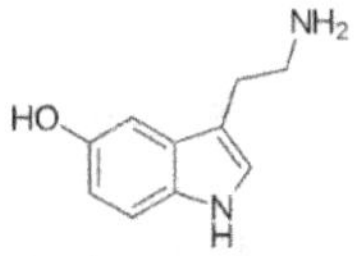

Serotonin

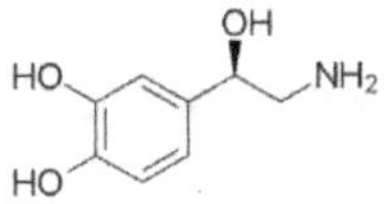

Norepinephrine

Dopamine is your brain's way of saying, "Hey, do that again." It spikes when you eat, when you win, when you fall in love. It drives motivation and the whole pleasure-and-reward cycle, and it's the system that addictive drugs like cocaine and meth (and yes, TikTok) crank into overdrive. Think of it as the chemical of desire.

Serotonin, also called *5-hydroxytryptamine*, or *5-HT,* is the mood regulator. Most of it (around 90%) actually lives in your gut, not your head, and it has a hand in your mood, your sleep, your appetite, your digestive movement, and a surprising amount of your social behavior. It's what SSRIs target, and it's heavily involved in psychedelics, which tend to mimic it or latch onto its receptors.

Norepinephrine, also known as *noradrenaline*, runs your stress response. It pushes up your heart rate and blood pressure and snaps you to attention. That's useful when there's actual danger, but keep it running high for too long and

you get anxiety and lousy sleep. It usually shows up alongside dopamine in stimulants.
We've met the main characters, the tiny chemical messengers that shape your thoughts, moods, and perceptions, but how do they actually do their work? Neurotransmitters don't just float around aimlessly. To change anything in the brain, they have to interact with something. That something is the receptor. It's the other half of the conversation.

Receptors: How Molecules Talk to Your Brain

Receptors are proteins sitting on the surface of your brain cells, each one shaped to fit a particular neurotransmitter like a key fits a lock. When the right chemical slots in, the receptor reacts, maybe by opening a channel, maybe by firing a signal off inside the cell. That one moment of binding kicks off the cascade that becomes a thought or a movement.

Your brain runs dozens of receptor types, each with its own job. Some only answer to your own natural chemicals. Others can be fooled, or outright hijacked, by drugs shaped just closely enough to fit the same locks.

For example:

- **5-HT receptors** are the primary receptors for serotonin. The one you'll hear about most in this book is **5-HT2A**, which is heavily involved in perception, emotion, and consciousness. This is the main target of psychedelics like LSD and psilocybin.

- **D1 and D2 receptors** respond to dopamine, the chemical of motivation and reward. Drugs like cocaine and methamphetamine stimulate these systems directly.
- **NMDA receptors** are part of the glutamate system, which helps with learning, memory, and awareness. Dissociatives like ketamine and nitrous oxide block these receptors, cutting off the brain's normal feedback loops.

When we talk about what a drug does, we're mostly talking about which receptors it binds and what happens once it's there. Picture the receptors as switches on your brain's control panel. Drugs flip them, and depending on which ones, you might feel euphoric, or confused, or strangely alert, or like you just met God. Every psychoactive drug works by either increasing, blocking, mimicking, or interfering with one or more of these chemicals or receptors:

- **Agonists** mimic neurotransmitters and activate receptors. (LSD mimics serotonin.)
- **Antagonists** block receptors, preventing them from activating. (Some antipsychotics block dopamine.)
- **Reuptake inhibitors** prevent recycling of neurotransmitters. (Such as cocaine.)
- **Releasers** force neurotransmitters into the synapse. (MDMA, MDA.)

When a drug floods a synapse, the rush can be euphoric or it can be overwhelming. Either way, the brain hates the chaos and pushes back. Receptors shut off. Signals weaken. Homeostasis kicks in.

That's the reason for comedowns, and for the "Suicide Tuesday" serotonin crash after a weekend of rolling. It's also why tolerance builds. Your brain is always hauling itself back toward balance, and knowing how it does that is the first step in not wrecking it.

What a Drug Is

A drug, at its simplest, is a molecule that interacts with your biology to produce change.

Some, like ibuprofen, knock down inflammation. Others, like caffeine, block your adenosine receptors and keep you awake. And some, like LSD or cocaine, go deeper, reaching into your neurochemistry in ways that bend consciousness itself.

LSD and psilocybin settle onto serotonin receptors and change how signal moves through your visual and emotional circuits. MDMA opens the floodgates on serotonin and dopamine at once, which is where the empathy and euphoria come from. Cocaine works on dopamine alone, blocking its reuptake so the reward signal just keeps hammering.

Drugs alter how your brain interprets reality itself. You're an observer sitting inside a complex system. Your body pulls input in through the senses and assembles it into a picture of the world in the cortex, with the thalamus doing the sorting in between. That picture gets filed away and checked against everything else you know, and if the story holds up, you believe it.

Psychedelics and dissociatives don't just "make you feel weird." They reach into the very process that builds your sense of identity and time. That's why a psychedelic can dissolve the ego, and why MDMA can hit you with a wave of love and trust you never asked for. Even casual use can quietly rewrite a memory, or the meaning you'd hung on it.

This is about altering the deepest code your brain runs on. One of the most important things about understanding your brain as a biochemical system is that it's changeable. Your thoughts can shape your brain. So can your behaviors. And yes, so can the substances you take.

It's called *neuroplasticity*, and it's the reason almost anything can reshape you over time: therapy, a meditation habit, a hard year, a psychedelic, even chronic stress. How you think and act is never fully locked in. Every new experience lays down a pathway, and every time you repeat it, that pathway gets stronger.

Used carefully, some substances seem to open that plasticity up. MDMA has shown up in trials as a tool for PTSD. Psilocybin is being studied for the anxiety that arrives at the end of life, and ketamine is already in use for depression that won't respond to anything else. But plasticity cuts both ways. A heavy trip can leave a mark you didn't see coming.

That's why set and setting matter so much, and why integration matters afterward. It's why you need to understand what you're walking into before you start. The more clearly you understand how your brain works, the more deliberately you can decide what to do with it.

Set & Setting – Packing for Your Trip

"She generally gave herself very good advice, though she very seldom followed it."

Every psychedelic journey is a collaboration between chemistry and context. The substance you take is only one part of the equation. The other parts are invisible, but no less powerful: your mindset, and the world around you. Psychonauts call this *set and setting*, and understanding it is as essential as understanding the molecules themselves.

A Brief History

The idea of "set and setting" might sound modern, but it has always been part of how humans approached altered states. Indigenous cultures rarely took mind-altering substances casually. There were rules, roles, and rituals around them. A shaman might fast for days. Songs were sung and the space purified before anyone began. Sacred mushrooms weren't taken at random, and ayahuasca wasn't passed around at a party. These experiences were shaped by boundaries, physical and cultural, that shaped what the participant would see and feel. You didn't just take the drug. You entered a ceremony.

It took Western science until the mid-twentieth century to start catching up. In the early 1960s, Timothy Leary, Ralph Metzner, and Richard Alpert (later known as Ram Dass) began using the phrase "set and setting" for the conditions that shaped a psychedelic journey. "Set" was the internal world: your mood, your expectations, the mental state and personality you walk in with. "Setting" was everything outside you, the physical space and the people in it and the emotional weather of the room.

Leary later wrote that the drug itself was only a catalyst. It opened the door, but what lay on the other side depended almost entirely on context. The same molecule could run through a clinical session, a jungle ceremony, or a warehouse rave and come out completely different each time. That became a cornerstone of psychedelic theory: these substances are non-specific amplifiers. What they bring up is shaped by what you bring in.

Even after psychedelics were pushed underground, the lesson held. Through the 1970s and 80s, underground therapists guided patients through sessions in secret, often paying meticulous attention to the music, the lighting, and the emotional tone of the space. Festival culture built its own version, handed from person to person: set your intention, find a safe place, stay with people you trust. If things go sideways, change the music or the lighting, or just step outside.

This wisdom survived through oral tradition, not institutional support. While governments tried to erase these substances from public life, people kept using them, and kept getting better at it. The core insight never changed: psychedelics are powerful, but the context you take them in is what decides what they mean.

Now that clinical research has come back and public interest has surged, set and setting is center stage again. Modern therapy rooms are designed to cut distraction and make people feel emotionally safe. Participants wear eyeshades and headphones. Guides stay close but quiet. Nothing is left to chance, not to control the trip, but to respect the conditions that shape it.

This is about intention. Every journey needs a map, and set and setting are how you draw it.

Set – The Landscape of Your Mind

"Set" is your inner world. It's the mindset you carry in: your mood, your thoughts, your habits of perception, the stories you tell yourself about who you are and what you expect to find. Psychedelics aren't passive tools. They don't simply show you "what's out there." They amplify what's already in there.

That includes your conscious mood, but also much deeper material. Past traumas. Quiet hopes. Lingering doubts. The parts of you that don't usually get airtime. Some of it may rise gently. Some of it may come through the surface with no warning at all. The trip doesn't care whether you planned to deal with it that day.

That's not a failure. It's the nature of these substances. They dissolve the usual filters and let you meet yourself raw. Sometimes that self is luminous. Sometimes it's confused or frightened. Often, it's several of those things at once.

Before you take anything, be honest for a minute. Where are you emotionally? What are you carrying in with you? Is there something you've been avoiding? Are you hoping this will fix you, or just hoping it'll be a night off from your own head? There's no wrong answer, but a false one has consequences. If you tell yourself you're doing deep spiritual work when really you want an escape from your pain, the experience may call your bluff.

A good way in is to set an intention. That doesn't mean scripting the trip. It means naming a direction. Maybe it's "I want to release some fear," or "I want to feel more connected," or even just "I want to have fun." Think of these as anchors rather than goals. They give you something to come back to when the landscape starts to shift.

Don't get too attached to it, though. An intention isn't a command. Insist on controlling the trip and it'll show you exactly how little control you have. The tighter you grip, the more it tends to pry your fingers loose. Paradoxically, the more you let go of what you expected, the better your odds of finding something that actually means something.

Tending your set also means tending your mind in the days before. Don't doomscroll right before you dose. Don't watch a horror movie the night before, pick a fight with your partner, or spiral into existential dread on Reddit at 2 a.m. What you feed your nervous system in the run-up matters.

Instead, slow down for a few days. Get outside, cook something, stretch, sleep enough. Read things that leave you calm or curious instead of wired. Talk to someone you trust. Take a walk and ask yourself what you're actually hoping to find, or what you're willing to see.

Meditation helps here, even if you're bad at it. You don't have to sit cross-legged or chant. Close your eyes and breathe slowly for a few minutes, and try not to grab onto thoughts as they drift past. Focus on observing the moment. Just notice what comes up. You're not chasing enlightenment. You're practicing noticing your own mind, which is exactly the skill you'll want during the trip.

Set also includes your physical state. If you are sick, sleep-deprived, injured or overheated, consider postponing your trip. What was a minor annoyance when sober can be the catalyst for a bad time mid-trip. Similarly, if you are coming down from other depleting substances, such as alcohol, cocaine, or MDMA, this exhaustion will reflect in your experience. You don't need to be perfect, just stable.

Diet can matter too. Some people fast, some eat light, some make a favorite meal in advance. There's no single rule beyond listening to your body. Greasy food or too many stimulants can spike anxiety or turn your stomach, and it's worth avoiding aged or fermented foods in particular. Treat your body like a vessel, not a garbage disposal.

If it's your first time, it helps to talk to someone who's done this before and say your fears out loud. Name them. "I'm afraid

I'll lose control." "I'm afraid I won't come back." Most fears like these are rooted in myth more than chemistry, but saying them makes them easier to meet if they actually show up.

Your set includes your story, too, the beliefs you carry about what psychedelics even are. Maybe you hold them as sacred sacraments, or see them as tools for working through trauma, or figure they're just fun. Whatever it is, that belief is part of your lens, and it shapes what you notice and what you skip right over.

You don't have to drop your story, but you should see it clearly. You're not showing up as a blank slate. You're showing up with a framework, and that framework is going to get tested. Sometimes it deepens. Sometimes it falls apart. Either way, it helps to know what you brought.

None of this is about being perfectly prepared. Perfection isn't the goal; room to breathe is. An inner world that's too tightly wound and overcrowded won't have space for anything new. Clear a little of it, and the trip may fill the space with something you didn't see coming.

Your set is the psychological soil the experience grows in. You don't have to control what comes up. You just have to make sure the ground is ready.

Setting – The World Around You

"Setting" is your external world: the room, the light, the music, the people around you. Psychedelics turn up sensory perception and emotional resonance, so what was background before the trip can become foreground during it. A flickering bulb, an odd painting on the wall, a sudden noise outside, any of it can swing the whole emotional tone.

A lot of people like to prep their space as part of a personal ritual. That might mean stringing up party lights, or burning sage, or laying out the things you'll want on hand. Plan for what you actually enjoy. Do you like something to fidget with, or to draw on, or room to dance? There's no correct answer, only the setup that lets you let go.

Many people experiment with psychedelics at concerts or festivals; while there's nothing inherently wrong with this, these tend to be the hardest environments to control, especially if you're not experienced with psychedelics. Make sure you have somewhere to go if things get overwhelming, and have a safe place you can be for the duration of the trip. Have a sober sitter who can arrange travel, if needed, and who can respond to your needs.

Keep a drink within reach, and snacks if you think you'll want them. Make sure the bathroom is easy to get to. Turn your phone off, or at least silence it and put it somewhere you can't reach without effort. Sort out music ahead of time, or a gentle way to change it. Wordless tracks usually work best, though some people like vocals or rhythms or the sound of rain.

Objects help. A favorite stuffed animal, a smooth stone, a photo you love, any of these can pull you back if you start to drift.

Some people like incense or essential oils, though those can get overwhelming fast if you're sensitive to smell. Be deliberate about your sensory palette.

Steer clear of places that are chaotic or unpredictable, or full of people you don't trust, because the crowd is part of the setting too. On a substance that cracks your defenses open, you need to know the people around you actually have your back. Psychedelics can leave you vulnerable in ways that are hard to predict. A wave of love or fear or confusion can hit, and the wrong crowd can turn that into something harmful. Pick the people who'll guard your boundaries instead of pushing on them.

Consent – The Social Contract

Psychedelics lower boundaries and heighten vulnerability. Emotions run deeper. Touch feels more intense. Words carry extra weight. This makes consent more important than ever.

Talk beforehand with the others who will be with you for the experience. Ask and answer clearly: Are we here to share openly or journey silently? Is touch okay? Are hugs okay? What are our physical boundaries? Is this a private journey or a communal one? What happens if someone needs space? What happens if someone spirals?

Consent should be unambiguous, and it must be ongoing. Just because someone agreed to something before the trip does not mean it's still okay during the peak. If you're unsure, ask. If someone withdraws, respect it. Consent must be explicit and enthusiastic. No means no. Silence means no. Confusion means stop, and being high never means yes.

If someone is in a state beyond where they could give sober consent, that also means no. This is why discussing consent beforehand is so important. Psychedelics can open doors and lower boundaries in ways you might not expect. You might feel cuddly, huggy, or sexually uninhibited. Discuss in advance what your boundaries are, and what choices your sober self would like for yourself when you can no longer make logical decisions.

Respect the decisions that others have made, too. When in doubt, ask. Don't assume your partner wants to talk. Don't assume your friend wants a hug. Don't assume shared vulnerability is the same thing as intimacy. Connection under the influence can feel profound; that doesn't make it romantic, sexual, or mutual.

If you're unsure whether an action is welcome, the best rule is simple: don't. Trust is fragile. If you want the experience to be meaningful, protect the space it happens in. That includes the emotional space of everyone involved.

Trip Sitters – Holding the Space

A trip sitter isn't there to guide, teach, or babysit. They're there to stay calm while someone else goes deep. Their role is simple, but important. When the walls of reality start to blur, a steady anchor can make the difference between feeling safe and feeling lost.

Psychedelics are unpredictable. Even in ideal conditions, they can bring up fear, grief, confusion, or overwhelming beauty. A trip sitter doesn't need to fix any of that. They only need to be present. Their calm nervous system speaks directly to yours. It says without words: *you are not alone.*

Trip sitting is especially important for first-time users, for high doses, or for anyone navigating trauma, anxiety, or unfamiliar territory. Even seasoned psychonauts sometimes choose to have a sitter nearby, not because they expect trouble, but because they respect the power of the experience.

The best sitters know how to stay centered. They don't get reactive or pour their own feelings into the space. They listen more than they talk, they watch without staring, and they stay available without hovering. The aim isn't to shape the trip. It's to hold the perimeter so the traveler can go all the way in.

If you are sitting for someone else, talk beforehand. Know what they're taking, and how much. Ask about their intention. Ask what kind of support they might want. Do they prefer silence or gentle check-ins? Do they want to be left alone unless they ask for help? Is physical touch okay if they are in distress? What does grounding look like for them? Water, a hand to hold, a change of music, a walk outside? Agree on signals and limits, then trust the agreement.

Have supplies ready: water, snacks, blankets, extra layers, a bathroom path that's easy to navigate. You don't need to prepare for every possible outcome. You only need to be prepared to remain calm if something unexpected comes up.

Avoid leading questions or interpretations. If they want to talk, let them. If they want to be quiet, let them, and don't project your own meaning onto their silence. Don't try to explain their trip to them. Your job isn't to write their story; it's to make sure they have the space to tell it themselves.

If they enter a difficult state, such as crying, looping, or becoming frightened, stay close and speak gently. Remind them where they are, that they are on a substance, and that it will pass. Offer grounding: "You're safe. You're at home. I'm here with you." Help them reconnect to their body; gentle stretching helps. Suggest breathing slowly. Offer water, a blanket, a hand to hold, or a different sound or space.

If things escalate beyond your comfort level, you should have a plan in place. Who can you call? What steps are agreed upon? You are not expected to be a therapist or a medical professional.

Your full attention is part of the act of sitting. Don't multitask. This is not the time to scroll your phone or work on emails in the background. It doesn't mean you need to stare or hover. Just being nearby, or even in the room, is usually enough. If the journeyer needs you, they should know that you are still there with them.

If you are serving as a trip sitter, you need to be sober. This is not optional. The person in your care is in a heightened and vulnerable state. Your ability to offer steady support depends entirely on your clarity and judgment. Even mild intoxication can make you less responsive, less attuned, or more likely to miss signs of distress. You're not there to vibe along or "guide the trip"; you're there to anchor the experience so someone else can let go and explore freely.

Being sober means you can respond to real needs in real time, whether that's refilling water, changing the music, or recognizing when something has shifted. You need to be able to discern if the person needs reassurance, grounding, or even medical attention. It's not a backstage pass to a psychedelic journey. It's a responsibility, and it should be taken seriously.

Sitting is an act of service. It asks for humility, patience, and attention. It is not glamorous. You may be quiet for hours. You may feel like nothing is happening. Ultimately, though, your presence creates the safety that allows real change to happen. You are holding the edges so the center can expand.

If It Starts to Go Sideways – Responding to a Bad Trip

First thing to remember: this, too, shall pass. Psychedelic states move in waves, and what feels like a crisis right now can melt into calm a few minutes later. The brain is flexible. No trip lasts forever. Breathe.

Change the environment. It's the most reliable lever you've got. Psychedelics amplify whatever's coming in, so the fastest way to change what's happening in someone's head is to change what's around them. Inside, open a window, dim the lights, turn the music down or off; outside, move somewhere quieter and less crowded. Even a small change in temperature or sound or light can reset the whole trajectory.

Aim for the opposite of where they are. If they're over-stimulated, pull them toward calm: a dim room, soft fabric, a familiar voice, a hand to hold. If they're sunk in their own head, looping through something dark or self-critical, pull them outward. Cartoons work better than you'd expect, and so does anything concrete enough to give the mind a place to land. Tactile anchors help too. Fidget toys, smooth stones, a textured blanket, even ice in a glass of water. The body needs reminders that it still has edges, and a real surface against the skin gives it one.

Watch for overshoot, though. Music, lights, location, and a new face all changing in the same minute can read as chaos even when each shift was meant to help. Move one thing at a time and give them a beat to settle in before reaching for the next.

Speak gently and stay grounded. If you're helping someone else, don't try to argue them out of fear with logic. Just reassure them that they're safe, that they took a substance, that the feeling is temporary, and that you'll be right there until it lifts. Sit with them. Offer water, a blanket, a hug, a quiet corner to themselves. Let them set the pace.

Don't hover or ask "are you okay?" every couple of minutes. Be cool. They don't need reminding that they're in distress. The more you act like something is seriously wrong, the more they'll catch it from you. Nothing escalates a bad trip faster than a room full of people treating it like one. Questions like "Are you still having a bad trip?" or crowds of people rushing to check on them will just reinforce that they're having a hard time. If someone is struggling, gently remove them from the situation and try not to call attention to them.

Often, a bad trip takes the form of self-critical feelings or fixation on negative thoughts or memories. It's not your place to interpret them or offer judgment. People in psychedelic distress will often express raw emotions in ways that they wouldn't when sober. They may cry, shout, curse, or laugh. This isn't a time to take offense or pick a fight. Be kind.

If they're stuck in a frightening or self-critical thought loop, try to interrupt it gently. A hand to hold, a change of music, a simple question that pulls them into the present: "what do you think of this song?" or "what's your favorite thing in this room?" That's not just distraction. It's anchoring.

Avoid mirrors. Avoid crowds. Avoid judgment. Don't laugh at someone's distress, and don't tell them to "just enjoy it." Don't force movement or conversation.

If you're the one having the hard time, remind yourself this is chemistry, not a permanent state. You are not broken. You are not dying. You are tripping. This will pass like everything else. Try lying down, breathing slow, holding onto something familiar. Let go of the need to steer. Surrender isn't giving up. It's letting the storm blow through without fighting it.

If it feels truly unmanageable and you planned ahead, you might have a benzodiazepine such as Xanax on hand as a "parachute" (*if prescribed and available*). Use it as a genuine last resort, and never with alcohol. These end the trip, sometimes abruptly, so if you reach for one, do it with intention rather than panic.

Afterward, give yourself space. Talk about what happened, write it down, take a walk, let it settle. A bad trip doesn't mean you failed. Sometimes it's your psyche showing you something faster than you were ready for, and sometimes it's just noise. Either way, you came through it.

Psychedelics are amplifiers. Respect the volume knob. And if you ever feel lost in one, hold onto this: you are not alone, and this is part of the journey too.

In Case of ACTUAL Emergency

Signs of serotonin syndrome (such as overheating, uncontrollable tremors, or seizures), or signs of opioid overdose (blue or grey skin tone, choking, gurgling, or unexpected loss of consciousness) mean that you need to take action NOW.

If a true medical emergency occurs, such as seizure, heart attack, stroke, severe injury, or signs of opioid or stimulant overdose, **do the right thing and call 911** or have a completely sober person drive the affected person(s) to the hospital or emergency room.

If you suspect opioid overdose and have Naloxone (Narcan), administer it per the manufacturer's directions. It won't hurt someone who doesn't need it.

Keep a cool head. Remember: Emergency responders aren't there to get you in trouble; they just want to help save a life. Be honest with them. Make sure you're able to share how much of what was taken, and when. Estimate if necessary. If you think adulterants may be the issue, save a small sample.

All 50 US states have "Good Samaritan" laws that protect individuals helping others in distress from legal prosecution, often requiring they stay on the scene, give their real names, and cooperate with responders. Exact requirements and protections vary from state to state. Contact a legal professional for advice and learn the laws in your area.

Integration – What Happens After

A trip doesn't end when the substance wears off. Some of the most important processing happens in the hours, days, and weeks that follow, when you're making sense of what you saw, felt, and understood. This is called integration, and it matters as much as set and setting. Without it, the insights dissolve into fragments. You're left with a vivid memory and confused feelings, but nothing that sticks.

Integration means taking time to reflect on what the experience showed you, even if it's hard to put into words. Journaling, drawing, or just sitting with your thoughts can help. Some people prefer to talk with a trusted friend or therapist. You don't need to turn every trip into homework. You just need to create space for meaning to settle in. Psychedelics can show you possibilities, but they won't carry those possibilities into your life for you. That part is yours.

Give yourself room to land. After a powerful trip, your nervous system needs calm and care. Plan a day or two where you don't have to rush anywhere or make big decisions. Spend time outside, where your senses can reset without the constant pull of screens and noise. Simple grounding practices like walking barefoot, stretching, or breathing deeply can help your mind settle and your body feel safe again. Meditation, even just five minutes of sitting with your breath, can keep the insights from slipping away too quickly. Eat nourishing food, sleep as much as your body asks for, and resist the urge to immediately jump back into the grind. Integration happens over days and weeks, not all at once. It's the patient practice of turning fleeting clarity into something you can actually use.

The quality of your set and setting will often determine how easily you can integrate what happened. A chaotic or negative environment can leave you with confusion or even trauma, while a safe and intentional one can provide insights that remain clear and useful long after the trip ends. The preparation you do before shapes the material you have to work with afterward. Both deserve the same level of care.

Putting It All Together

You can't control everything, but you can influence how your journey unfolds. Prepare your mind and your space, know the people around you, set an intention, and give yourself time afterward to process what you find.

What you take matters, but how you enter the experience and what you bring back matter even more.

MDMA/MDA – "Ecstasy"/"Molly"

"I know who I WAS when I got up this morning, but I think I must have been changed several times since then."

MDMA does one thing no other substance quite replicates: it eliminates the distance between you and everyone else. The warmth and connection are real. The music suddenly sounds amazing. What's behind all of it is a flood of serotonin, dopamine, and norepinephrine so complete that for a few hours, your brain runs a version of itself it didn't know it had. Understanding what's actually happening in there is the difference between using it well and using it up.

A Brief History

Before MDMA ever saw a dance floor, it spent decades sitting in a chemical catalog. In 1912, a German chemist named Anton Köllisch, working for the pharmaceutical company Merck, first synthesized MDMA while searching for blood-clotting compounds that could help control bleeding during surgery. At the time, the compound's empathogenic properties were unknown, and after initial testing, it was shelved with little fanfare. For decades, it remained an obscure entry in Merck's chemical catalog, a curiosity rather than a breakthrough. It was not until the 1950s, when the U.S. Army conducted animal tests on MDMA as part of its EA-1475 chemical warfare program, that the molecule briefly resurfaced, though once again its true potential went unrecognized.

The real story begins in 1976, when Alexander "Sasha" Shulgin resynthesized MDMA in his Lafayette, California, lab and tested it on himself. Shulgin had spent his career cataloging novel psychoactive compounds in the notebooks that would become his famous book, *PiHKAL: A Chemical Love Story*. After experiencing MDMA himself, he recognized its unique ability to open emotional doors without overwhelming the mind. He introduced it to his circle of friends and psychotherapists, most notably Oakland psychologist Leo Zeff, who began using the compound as a therapeutic agent. Zeff was preparing to retire. He didn't. He spent the next eleven years introducing MDMA (which he called *Adam*) to thousands of therapists, who used it to reach material that talk therapy alone couldn't. The work was still legal and almost entirely underground.

By the early 1980s the same compound was legally sold in Dallas gay clubs under the name "*Ecstasy*." It made people feel close to each other, which was useful in a therapy session and even more so on a dance floor. As demand grew, underground chemists began producing MDMA in larger quantities, and its popularity spread across the United States and Europe. The drug was still legal, but its growing association with club culture and countercultural freedom drew the attention of the authorities. In 1985 the DEA moved to schedule MDMA. Therapists and researchers testified against it at the hearings. The administrative law judge who oversaw the process, Francis Young, recommended Schedule III. The DEA overrode him and placed MDMA in Schedule I anyway. That is the classification it still has, forty years later.

Scheduling didn't slow it down, but it did change what was in the pills.

Through the 1990s, rave culture ran on MDMA, first in the US and UK, and then the world. It was also increasingly not MDMA. Pills sold as Ecstasy started showing up cut with caffeine, methamphetamine, MDA, and eventually the deadly analogs PMA and PMMA. Deaths at raves made the news, and most of the news blamed the drug rather than the supply.

In 2003 Congress passed the RAVE Act (Reducing America's Vulnerability to Ecstasy), which made venue owners and promoters legally liable for drug use on their premises, whether or not they knew about it or had tried to prevent it. It was aimed at shutting down raves. What it actually did was punish the people trying to make raves safer. An owner who allowed a medical tent, a water station, or on-site drug checking was handing prosecutors evidence that he knew drugs were present. The tents and water stations went away. The drug use didn't. It moved somewhere with no one watching.

Underground therapists kept working through all of it. Zeff had died in 1988, but the network he built kept meeting. Rick Doblin founded MAPS in 1986 specifically to get MDMA-assisted therapy back through the FDA. It took thirty years to get to a Phase 3 trial.

The Phase 3 results, when they finally came, were striking. In the first of two MAPS trials for PTSD, published in 2021, 67% of participants who received MDMA-assisted therapy no longer met the criteria for PTSD at the two-month follow-up, compared to 32% in the placebo group. The effect size was roughly four times what existing PTSD medications produce. In 2017 the FDA had already granted the treatment Breakthrough Therapy designation. For a while it looked like approval was a matter of when.

It wasn't. In August 2024, the FDA rejected the Lykos Therapeutics application. The rejection cited concerns about the trial data and unblinding (participants can usually tell whether they got MDMA or placebo, which is a real methodological problem), and pointed to ethical violations at one Canadian trial site where a therapist had a sexual relationship with a participant. Two of the trial papers were later retracted. Lykos cut about three-quarters of its staff and started designing a third Phase 3. The work continues.

After more than a century of prohibition, moral panic, and quiet clinical promise, MDMA's fight for legitimacy is still very much underway.

How MDMA Works

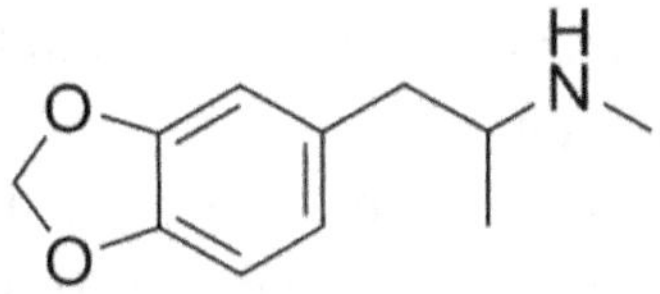

MDMA (3,4-Methylenedioxymethamphetamine)

MDMA is best known for its warm, empathic glow, but underneath the feeling is a dramatic neurochemical event. Technically, MDMA is a *serotonin-dopamine-norepinephrine releasing agent.* Unlike most antidepressants, which block reuptake and passively increase neurotransmitter levels, MDMA goes further. It actively hijacks the machinery that neurons use to recycle neurotransmitters and forces it to run in reverse.

In a typical neuron, neurotransmitters like serotonin are packaged into bubbles called *vesicles*, released into the synapse when needed, and then cleared away by transporter proteins. These transporter proteins, such as the serotonin transporter (SERT), act like vacuum cleaners, sucking up excess neurotransmitters to stop the signal and prepare the neuron for the next one. It's efficient. It keeps your brain from being overstimulated.

MDMA flips that system on its head. Once inside the neuron, MDMA enters the presynaptic terminal and binds to SERT. Instead of just blocking reuptake, it *reverses the direction of the transporter*. Like opening an emergency exit valve, it causes serotonin stored inside the neuron to be dumped out into the synapse, regardless of whether a signal was just sent. This flood isn't tied to experience or thought. It's completely chemical.

The same reversal occurs with dopamine and norepinephrine, though serotonin release is the most dramatic. The result is a synaptic space that's suddenly brimming with neuro-transmitters. Your emotional tone shifts. Empathy rises. Sensory input sharpens. Touch, music, and human connection feel amplified and significant. The entire brain becomes an echo chamber for meaning.

This transporter reversal also explains the time course of MDMA's effects. The rush begins about 30-60 minutes after ingestion, peaks in 1.5 to 2.5 hours, and then gradually tapers as the neurotransmitter stores are depleted. The body metabolizes MDMA primarily via the liver enzyme CYP2D6, which converts some of it into MDA, a longer-lasting, more psychedelic cousin that may account for late-stage visuals or a second "wave" in the roll.

Forcing the release of neurotransmitters isn't without consequences. Your brain doesn't have time to make more in real time. After the roll, levels of serotonin and dopamine are significantly reduced, and your receptors, saturated and over-whelmed, may downregulate to compensate. This is the biochemical basis of the comedown. The infamous "Suicide Tuesday" isn't just emotional; it's a depleted system trying to rebalance itself.

Over time, repeated use without recovery can exhaust this system further. That's why MDMA is not physically addictive in the traditional sense (it doesn't hook you the way opioids or stimulants do), but *psychologically reinforcing*. It offers a powerful emotional state that some users chase repeatedly, even as the magic fades. That fading, known as tolerance, is a sign your receptors are adapting, and your system needs rest.

The mechanism matters because everything you feel on MDMA was already there before you took it. The drug just brings it to the surface. What gets through is real, but it's concentrated in a way that ordinary experience isn't. Sunlight through a magnifying glass turns gentle warmth into something that can burn. How close you sit is up to you.

Smart Guidelines for MDMA

Rolling smart means being deliberate. Your neurons have limits, and you're the one who has to live with them tomorrow. Here are some guidelines worth keeping in mind:

Test your substances.

Use reagent kits like Marquis, Mecke, and Mandelin to screen for common adulterants. Simon's reagent can discriminate between MDA and MDMA. You can get a full test kit from DanceSafe or other reputable places online. Test for fentanyl even if you're "sure" it's clean. False confidence has ended lives.

Don't take 5-HTP during the roll.

It seems like it makes sense: MDMA depletes serotonin, 5-HTP replenishes it, but timing matters. Taking 5-HTP too soon can put you at risk for serotonin syndrome. MDMA releases so much serotonin that adding more is like throwing gasoline on a raging fire. Wait until the next day.

When people use 5-HTP, a common add-on is EGCg, green tea extract, or just strong green tea. Most of your body's serotonin is made in the gut, and your body is very good at converting 5-HTP out in the periphery before much of it ever reaches the brain. There's some evidence that EGCg can inhibit the enzyme that converts 5-HTP into serotonin, so it can slow that peripheral conversion and leave more 5-HTP available to cross into the brain. Get that serotonin where it will help!

Don't finger-dip/eyeball your dose.

MDMA is active in the 75-125 mg range. That's a tiny difference when you're scooping powder. Use a scale that measures to the milligram. They cost less than a half gram of mystery molly and can save your life.

Hydrate, but wisely.

Too much water can be as dangerous as too little. MDMA causes water retention and electrolyte imbalance. Sports drinks, such as Gatorade, Powerade, or Body Armor, are great alternatives to water, as they contain electrolytes and vitamins.

Go easy on the dosing.

You can always take more, but you can't take less. In clinical settings, the optimal dose was found to be a reasonable initial dose (80-100 mg) followed by half of the original (40-50 mg) two hours later. Especially if you're preloading on antacids, try starting low. Any more than that, and you're increasing the neurotoxicity and reaching diminishing returns on the effect. Taking more doesn't make you a hero. More drugs don't necessarily equal more fun. Try lower doses. Your brain will thank you for it later.

Space out your experiences.

Your serotonin system doesn't regenerate that fast. Back-to-back dosing increases the risk of long-term dysregulation, emotional burnout, and deeper comedowns. Wait at least a month. Three is better.

Mix with care.

Stimulants like cocaine, methamphetamine, or even strong energy drinks elevate heart rate and blood pressure, and can cause cardiac issues, even in people with no previous history of them. Stack them, and you stack the risk of cardiac arrest. It doesn't matter how healthy you feel. Your heart isn't made of titanium.

Mixing MAOIs and MDMA is a fast track to *serotonin syndrome*: a potentially fatal overload of neurotransmitters that can result in overheating, tremors, confusion, or seizures. Do not take MDMA within two weeks of MAOI use. This includes certain uncommon antidepressants (your doctor will tell you if you're being prescribed MAOIs), herbal supplements (like Syrian Rue or St. John's Wort), and even some diet aids.

Common SSRI antidepressants such as Lexapro, Celexa, or Prozac (or their respective generics) present a complicated interaction with MDMA. They tend to significantly dampen the effects of MDMA, but this is where the real danger lives. People who don't feel the roll will often take more, and that extra MDMA doesn't restore the magic. It just increases neurotoxicity and pushes you closer to serotonin syndrome. If you are on SSRIs, do not increase your dose to compensate. Take a reasonable amount, accept a gentler roll, and protect your brain.

Other common psychedelics, such as mushrooms or LSD, don't generally interact dangerously, but will amplify the effects of both substances. Lower your initial dose of both until you get a better feel for how they interact.

Lastly, DXM: Found in over-the-counter cough medicine and some adulterated pills, *dextromethorphan* (DXM) also impacts your serotonin levels and can increase the risk of serotonin toxicity when combined with other serotonergic substances. Serotonin toxicity can include hyperthermia; combine it with MDMA and you risk overheating dangerously fast.

Integrate afterward.

The journey doesn't end when the music stops. Reflect. Journal. Talk. Rebuild your biochemistry with food, rest, and time. Otherwise, the lessons of the night tend to dissolve by morning.

These guidelines exist because you deserve to be around long enough to grow from your experiences. Conscious use means paying attention, making choices on purpose, and staying connected to what's actually happening. It also means knowing when to sit one out. The molecules may be synthetic, but your life isn't, and it's worth protecting.

Vitamins: How to Hug Your Brain Back

Ask around and most people's recovery advice for MDMA sounds simple. Drink water, sleep it off, and you'll be fine in a day or two. That's not wrong. Water and sleep do most of the heavy lifting. But "fine in a day or two" skips over what your brain is actually doing while it recovers, and a little chemistry makes those days noticeably easier.

The warmth you feel on MDMA runs through your serotonin system, and serotonin is a limited resource. When you roll, you spend it faster than your brain can restock, and you put a real metabolic load on your mitochondria while you're at it. You are borrowing tomorrow's energy and tomorrow's good mood to pay for tonight. The supplements in this chapter are how you cover some of that cost up front and settle the rest afterward.

The flood of neurotransmitters MDMA releases is metabolically expensive, and your neurons burn fuel hard to keep firing. Burn fuel that fast and you create chemical byproducts that have to be cleared. When the cleanup can't keep up with the mess, you get *oxidative stress*. Picture an engine run in the red for eight hours straight, and you have the right idea.

A lot of that byproduct comes in the form of unstable molecules called *free radicals*. They're reactive, and left alone they damage whatever cell parts they bump into. Your body runs a standing cleanup crew for exactly this: antioxidants, which neutralize free radicals before they cause trouble. On a normal day the crew keeps pace. During a roll the mess outruns them, and that backlog is a good part of what the tail end of the night, and the next morning, actually feel like.

This means the comedown is not only in your head. The brain fog and the low, flattened mood are partly physical, the impact of your body doing repair work against the clock. Antioxidants won't change the peak. They aren't there to make the high stronger or softer. Their whole job is the cleanup, so you land back at baseline with less to haul out later.

Cleanup is only half the story. After a night of dumping neurotransmitters, your brain is both depleted and inflamed. Depleted means the raw materials are gone and have to be replaced. Inflamed means the tissue needs time and a calmer chemical environment before it settles. The supplements handle both sides: the building blocks to restock, and the antioxidant and mitochondrial support to bring the inflammation down. Do that on purpose and the roll winds down instead of crashing.

This protocol grew out of years of trial and error in the field, not a clinical trial. People who took specific supplements before and after rolling kept reporting the same thing: gentler crashes, and heads that felt clearer the next day. Those notes got passed by word of mouth and traded on online forums. The biochemistry behind each item, however, is real. The stack as a whole is field-tested, not FDA-approved.

The stack is built around timing, because your brain doesn't hit all its problems at once. What helps you six hours out is doing a different job than what helps you at the peak, and both are different again from what you want in the recovery days. Get the item right but the timing wrong and, best case, you've wasted it. With something like 5-HTP, worst case, you make things worse. The timing windows aren't trivial. They matter as much as what goes in them.

The goal is to protect your neurons from the oxidative wear, and give your brain enough support to hold on to whatever the night was actually about. You do not need to start with every item on this list, but knowing what each one does, and why it lives in the window it lives in, is what lets you build a version that fits your body and your budget.

Note that the vitamins on the following pages that I consider the most essential are marked with a star (★).

Six Hours Before Dosing

Start early. You're preparing your system for a high-energy, high-emotion event. Think of this as preloading the defense systems before the fireworks begin.

★ ***Magnesium glycinate (200 mg):***

A muscle relaxant that helps reduce jaw tension, teeth grinding, and eye twitching. Glycinate is the preferred form. It's easy on the stomach and calming without sedating. Don't get magnesium citrate or oxide: they have poor absorption and tend to have a laxative effect. You'll be taking lots of magnesium throughout the night, so get a form you enjoy taking. Don't get giant horse-pills that you hate taking and will abandon mid-roll. Magnesium glycinate also comes in easier-to-take forms, including gummies and chewables.

★ ***Sodium R-Alpha Lipoic Acid (Na-R-ALA, 100 mg):***

A potent antioxidant that helps prevent oxidative damage caused by neurotransmitter release. It supports mitochondrial function and reduces the cellular stress of rolling.

Look for ***Na-R***-ALA (Sodium-R-Alpha Lipoic Acid) instead of ALA (Alpha Lipoic Acid). The sodium form is more protective, since it isn't broken down as quickly. If all you have is ALA, it'll do, in a pinch, but you'll need to increase it to 500-600 mg, and take it every few hours.

★ ***Ginger (≈3 grams):***

A natural anti-nausea agent that may also inhibit liver enzymes (CYP2D6 and CYP1A2) responsible for converting MDMA into MDA. This can help reduce the intensity of the comedown and avoid the heavier and more psychedelic edge of MDA.

If you're using fresh ginger (and don't mind the bite), aim for a piece the size of a quarter (US 25¢ coin). Some people prefer ginger candy, like Gin Gins or Chimes. Any real ginger will do: fresh, candied, ginger beer, or in a tea. Just make sure it's the real root, not flavoring. Note that most commercially available ginger ale contains no real ginger. If you don't like the flavor of ginger, encapsulated ginger extract is available and substitutes well.

Want a downloadable version of this vitamin stack to keep on your phone or share with friends?
Scan this QR Code.

alteredstatesbook.com/harmreductioncards

One to Three Hours Before Dosing

This is your main supplementation window. You're building biochemical resilience and preparing for the flood of neurotransmitters to come.

★ ***Magnesium glycinate (200 mg):***
Build your magnesium stores up, since you're preparing for a lot of activity!

★ ***Sodium R-Alpha Lipoic Acid (Na-R-ALA, 100 mg):***
Builds antioxidant protection ahead of the oxidative stress.

★ ***Fresh ginger (≈3 grams):***
Ginger is slightly acidic, helps settle your stomach, and may help prevent MDMA from converting into MDA.

L-Tyrosine (500 mg):
A precursor to dopamine and norepinephrine. Helps refill the tank after those neurotransmitters are released en masse. L-Tyrosine notably improves your appetite the next day, often solving the "crying in my breakfast" phenomenon, and allows you to generally recover faster with proper nutrition.

N-Acetylcysteine (NAC) (600 mg):
N-Acetylcysteine (NAC) is a precursor to *glutathione*, one of the strongest antioxidants. It brings down inflammation and brain fogginess.

Vitamin C (1000 mg):
An antioxidant that helps buffer free radical production and supports immune function. It may also reduce neuroinflammation from serotonin release.

Grape seed extract (100 mg):

Another antioxidant, particularly good at crossing the blood-brain barrier. It supports capillary health and reduces the oxidative load on neurons.

Acetyl-L-carnitine (ALCAR, 500 mg):

Helps protect mitochondria from oxidative damage, supporting cognitive clarity and physical stamina during long rolls. May also reduce mental fog afterward.

EGCg (green tea extract, 400 mg):

Has multiple roles: antioxidant support, anti-inflammatory effects, and urinary relaxation (i.e., helping to counteract the notorious MDMA "can't pee" problem).

CoQ10 (100 mg):

Supports cellular energy production and protects brain tissue. Particularly useful if you're prone to headaches or fatigue after rolling.

Thirty Minutes Before

Tums or Rolaids (2 tablets):

These antacids reduce stomach acidity and can improve the oral bioavailability of MDMA. This means you can take less, and roll harder, with less of a hangover.

Be careful! If you are taking antacids beforehand, make sure to reduce your dose. Start with a 20% reduction (80-100 mg is a good start), and see how you feel. You can always take more, but you can't take less!

Any over-the-counter calcium carbonate antacid such as Tums or Rolaids will work. (Bonus points for the extra strength kind.) This doesn't need to be name-brand; the store brand is fine. Note that Prilosec, Pepcid, Nexium, or modern generation acid-reflux medications aren't what we're looking for here. They change the way your body produces stomach acid, but won't reduce the actual acidity. Stick to calcium carbonate-based antacids. Check the label if you're unsure.

During the Roll (Every Two Hours)

You're supporting the system while it's working overtime. Set an alarm on your phone if you're prone to forgetting.

★ ***Magnesium glycinate (200 mg):***
Helps calm muscular tension, jaw tightness, eye twitching, and muscle spasms common with long rolls.

★ ***Na-R-ALA (100 mg):***
Reinforces antioxidant protection to keep oxidative stress in check as neurotransmitter activity peaks.

Grape seed extract (100 mg):
Continues to scavenge free radicals and support vascular health, particularly important when heart rate and blood pressure are elevated.

N-Acetylcysteine (NAC) (600 mg):
Helps clean up any lingering neurotransmitters and reduces inflammation.

After the Experience (Same Night)

The chemical high is over, but your brain is still catching up. Sleep is critical, yet challenging.

Melatonin (5-10 mg):

A higher-than-usual dose may help restore circadian rhythm and reduce sleep disturbances common after MDMA. Melatonin has also been shown to have antioxidant effects, and acts as a neuroprotectant during serotonin depletion.

This seems like a lot of melatonin, especially since many of us are used to taking 0.5 to 3 mg falling asleep instantly. In normally operating brains, melatonin is produced directly from serotonin. At the end of a roll, your serotonin stores are depleted, leaving your body little raw material to produce melatonin. This contributes to the difficulty sleeping that many people experience afterward. Start with 5 mg and go up to 10 mg if needed. This higher dose will assist with overcoming this deficiency and help end the roll.

Hydration:

Drink when you feel thirsty, but don't overdo it. Electrolyte drinks (not plain water) help prevent hyponatremia and support neural recovery. Sports drinks, such as Gatorade, Powerade, and Body Armor, also often contain vitamins.

Next 2-3 Days: Recovery Phase

Your serotonin system is depleted. These days matter just as much as the party. Give your brain the raw materials it needs to recover. Try taking these before bed:

5-HTP (100-200 mg):
A direct precursor to serotonin. Only take it after the MDMA has fully cleared your system, typically 12 to 24 hours after your last dose. Taken too soon, it can worsen serotonin toxicity.

EGCg (400 mg):
Helps convert 5-HTP safely, reduces side effects, and enhances absorption. Always take EGCg with 5-HTP!

Vitamin B-6 (25-50 mg):
Acts as a co-factor for serotonin synthesis. Without it, your body can't efficiently turn 5-HTP into usable neurotransmitters. This can also give you wild and vivid dreams.

Magnesium glycinate (200 mg):
Continued support for restful sleep and muscle relaxation.

Melatonin (1-5 mg):
Continuing to support circadian rhythm and reduce sleep disturbances common after MDMA.

Hydration:
Continue the sports drinks (electrolytes!) over the next few days to flush out any remaining toxins and support recovery. Drink enough that your urine is clear.

Nutrient-Dense Foods

Over the next few days, try to eat a diet high in eggs (for choline and tryptophan), salmon (for omega-3s), oats and bananas (for tryptophan), and pumpkin seeds (for zinc and magnesium). These help rebuild neurotransmitter stores naturally.

From a recovery standpoint, the best breakfast you can make after a night of rolling is cheese and eggs. Containing much of what your body needs to replace those lost transmitters, it will help you bounce back fast. Make them a little saltier than usual; it'll help them be a bit more palatable. Add some orange juice, and it's a full meal!

What MDMA Offers

Used with care, MDMA offers more than euphoria. It can unlock empathy, soften trauma, and create a rare clarity in which emotional truths become speakable. In therapeutic contexts, it can help people reconnect with parts of themselves that have been buried or shut down. In social settings, it can turn strangers into confidants, but the insights it reveals are only as lasting as the intention you bring to them. Without respect for dosage, recovery, and timing, MDMA can leave you wrung out instead of transformed. If you approach it as a student (and not as just a consumer), it has the potential to change more than your night; it can change ***you***.

MDA/"Sass" – The Forgotten Cousin

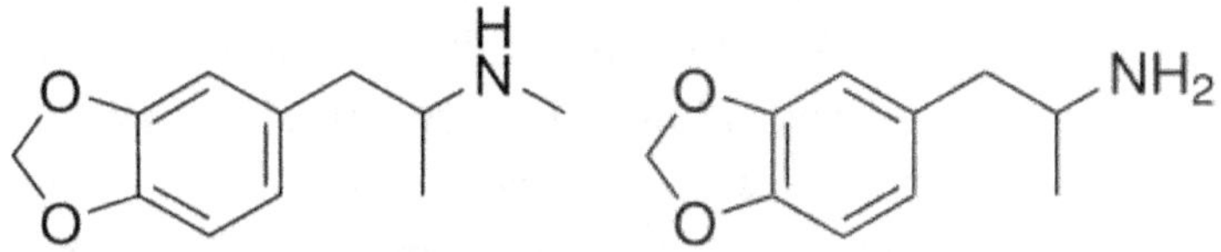

MDMA (left) and MDA (right)

If MDMA is the golden child of the empathogens, known for its warmth, openness, and therapeutic glow, then MDA is the moody older cousin. Stronger, longer, more psychedelic, and less predictable. It rarely gets the spotlight, but it's been part of the story from the beginning.

Like MDMA, MDA belongs to the phenethylamine family, a class of molecules that blend stimulant and psychedelic properties. Like MDMA, it releases serotonin, dopamine, and norepinephrine. MDA, however, leans harder into its psychedelic lineage. Its effects tend to be more visual, more body-heavy, and less socially connective. Some users describe MDA as a bridge between MDMA and LSD, and find it less euphoric, more introspective, and occasionally more challenging.

Experientially, MDA comes on slower and lasts longer. Where MDMA typically peaks between 1.5 and 2.5 hours after ingestion and fades within six hours, MDA can stretch past eight hours and often arrives in waves. People report intense visuals, vibrating patterns, shifting light, and a dreamlike quality to thought and emotion. At higher doses, it may dissolve the sense of self entirely, producing ego loss, cosmic unity, or, in the wrong setting, paranoia and confusion.

The body load is heavier. Jaw clenching, eye twitching, temperature swings, and gastrointestinal discomfort are common. MDA is also more likely to cause auditory distortions, closed-eye visuals, and a sensation of energy coursing through the body. It's often called *sass* in some circles, and it's definitely the "sassier" molecule; harder to dance with, but full of secrets if you're willing to listen.

How It Works

MDA, like MDMA, releases massive amounts of serotonin, dopamine, and norepinephrine, flooding your brain with neurotransmitters associated with mood, reward, and arousal. It also has a stronger binding affinity at 5-HT2A receptors, the same sites targeted by classical psychedelics like LSD and psilocybin. This dual action explains its hybrid nature. It's both stimulating like a party drug and expansive like a trip.

MDA isn't just its own molecule. It's also a metabolite. When you take MDMA, part of it gets broken down in your liver into MDA. This is one reason some people report a "second wind" or a mood shift midway through their roll when taking MDMA; it's MDA taking the wheel.

This is also why the comedown from MDMA can feel strangely different than the peak. MDA is less forgiving. Its neurotoxicity is slightly higher, and its aftereffects (such as fatigue, serotonin depletion, and melancholy) can last longer. Some people love the visuals and depth MDA brings; others find it cold, isolating, or difficult to integrate.

Some take ginger before and during the roll, not just to settle the stomach, but to inhibit the liver enzymes that convert MDMA into MDA. The theory is that by slowing this metabolic pathway (specifically CYP1A2 and CYP2D6 enzymes), ginger may help the roll remain more empathogenic and less psychedelic. While this mechanism hasn't been widely studied in clinical settings, anecdotal reports suggest that ginger can reduce the likelihood of a harder, more visual "second wave" that comes with MDA conversion.

This practice is far from foolproof, but it reflects an evolving awareness of how nutrition, metabolism, and pharmacology intersect. Like all things in this space, the key is to test, observe, and listen to your body.

Common Misconceptions

Because most people encounter MDMA in pressed pills or powders labeled "Ecstasy" or "Molly," they may have unknowingly taken MDA. A particularly long, wavy, or visual roll might not be a strong dose of MDMA, but a different compound entirely.

Many recipes to create MDMA involve MDA as a precursor, and some chemists will purposely leave a mix of MDA and MDMA in their final product. This has given rise to urban legends that some pills are "heroin-based" or "speedy." In truth, most of this confusion comes from different ratios of MDA and MDMA, rather than other drugs as adulterants. Only proper reagent testing (using Simon's reagent) can tell the difference between MDMA, MDA, and something you didn't want.

Don't guess. Test.

Harm Reduction and Recovery

Because MDA is longer-lasting and more stimulating, it puts more stress on your system than MDMA. Hydration, electrolyte balance, and body temperature regulation are essential. Preloading with magnesium can help reduce jaw tension and muscle cramps. Ginger may help with nausea, but its usefulness doesn't stop there.

The same nutritional stacks, schedules, and warnings that apply to MDMA also apply to MDA. Note that MDA is stronger than its cousin and lasts longer, making supplementation during the roll that much more important. Lower your dose from what you'd take of MDMA; you can always take more, but you can't take less.

Afterward, give yourself space to reintegrate. MDA tends to leave a heavier emotional footprint than MDMA. It's less the feeling of a hug, and more the memory of an intense conversation you're still processing. Prioritize rest, quiet reflection, and nourishing, high-tryptophan foods like eggs, salmon, oats, or sunflower seeds. These support serotonin rebalancing and emotional stability.

Wait at least a month between MDA sessions, preferably three or four. Combining MDA with other stimulants, psychedelics, or psychiatric medications (especially MAOIs, SSRIs, or antipsychotics) can dramatically increase the risk of serotonin syndrome or neurotoxicity. Know your combinations. Know your limits.

What It Offers (and Demands)

MDA is harder to love than MDMA. It's longer, heavier, more visual, and less forgiving. The warmth is there, but it's buried under layers of introspection that can feel more like excavation than euphoria. Where MDMA opens the heart and makes the room feel safe, MDA is more likely to open a door you forgot was there and show you what's behind it. That can be sincerely valuable. It can also be sincerely rough.

Some psychonauts seek MDA out for exactly that reason. They're not looking for connection or comfort. They want the raw edges, the unfiltered view, the version of the experience that doesn't smooth anything out for you. MDA has a reputation in those circles as a teacher with no patience for small talk. It shows you what it shows you, and it doesn't particularly care whether you were ready.

That's also why MDA demands more preparation, more recovery time, and more honest self-assessment than its cousin. The body load is real. The emotional aftermath can linger for days. MDA can leave you feeling like you've been through something significant but can't quite name it yet, and that ambiguity takes time to resolve. If MDMA leaves you glowing, MDA is more likely to leave you thinking.

For those who approach it with respect, knowledge, and enough humility to sit with discomfort, MDA can crack open insights that MDMA's warmth might never reach. It's a different tool for a different kind of work. Just know what you're signing up for before you take it.

Cannabis/THC – "Weed"

"Somehow it seems to fill my head with ideas — only I don't exactly know what they are."

The One Everyone Thinks They Know

Cannabis is probably the first substance many of you tried, and the one you've used most since. It's also the one least likely to appear in a harm reduction conversation, because somewhere along the way, "it's just weed" became a complete sentence.

Mostly, that's fair. Cannabis has a favorable safety profile compared to almost everything else in this book. Nobody has ever died from a THC overdose. It doesn't create cocaethylene in your liver. It won't fry your serotonin system or punch a hole in your nasal septum.

But "practically harmless" isn't the same as "no harm reduction required," and that assumption has put a lot of people in very unpleasant situations. The edible that didn't kick in for ninety minutes. The bad trip that went sideways the moment someone passed a joint. The synthetic cannabinoid that turned out to be nothing like what it was sold as.

Cannabis is the most widely used psychoactive substance at festivals, concerts, and gatherings worldwide. That ubiquity is exactly why it deserves a chapter. Not because it will kill you, but because most people using it have never thought about it critically, and because the landscape has changed significantly in the last decade. Today's cannabis is not your parents' cannabis, or even the cannabis of ten years ago.

A Brief History

Cannabis has been with humans for a long time. Archaeological evidence points to cultivation in Central Asia as far back as 12,000 years ago, making it one of the oldest cultivated plants on record. Ancient Chinese texts from around 2700 BCE reference it for pain relief and inflammation. In India, it has been woven into Hindu spiritual practice for millennia, particularly in the form of *bhang*, consumed during festivals like Holi. The Scythians burned it in enclosed spaces as part of burial ceremonies. Herodotus documented this in the 5th century BCE with barely concealed astonishment.

Western medicine's formal relationship with cannabis began in 1839, when an Irish physician named William Brooke O'Shaughnessy, working for the East India Company in Calcutta, published a landmark paper documenting its medical applications. He had observed Indian physicians using cannabis preparations for pain, muscle spasms, and convulsions, conducted his own trials, and shipped samples back to Europe. By the 1840s, cannabis tinctures were being prescribed by physicians in Britain and the United States for everything from migraines to menstrual cramps to the tremors of Parkinson's disease. Queen Victoria's personal physician, Sir J.R. Reynolds, reportedly prescribed it to her for menstrual pain. For the latter half of the 19th century, cannabis was a standard and unremarkable fixture of Western medicine.

It arrived in the Americas through colonial trade routes, and the 19th-century patent medicine industry embraced it enthusiastically. Cannabis preparations appeared in catalogs alongside laudanum and cocaine tonics, marketed for pain, insomnia, and what Victorian-era physicians broadly categorized as "nervousness." It was unremarkable, legal, and widely available.

That started to change in the early 20th century, not because of science, but because of politics. The 1910s and 1920s brought waves of Mexican immigration into the American Southwest, driven by the upheaval of the Mexican Revolution. Cannabis use came with those communities, and it already had a name in Mexican Spanish: *marijuana*. The plant had existed in American pharmacies for decades under the name cannabis, but marijuana was foreign-sounding, and that was the point.

The same era saw cannabis move into Black communities in New Orleans, where it became associated with jazz culture, late-night venues, and a social world that mainstream white America viewed with a mix of fascination and contempt. Louis Armstrong was a devoted and unapologetic smoker his entire adult life. The association between cannabis, Black musicians, and integrated social spaces made it a convenient target.

Then came Harry Anslinger. Appointed as the first commissioner of the Federal Bureau of Narcotics in 1930, Anslinger built his career on the criminalization of cannabis, and he did it largely through racism and manufactured panic. His public campaign linked cannabis to violence, sexual deviance, and moral collapse, drawing explicitly on racial anxieties about Black and Mexican men. He promoted "reefer madness" as a governing philosophy. The Marihuana Tax Act of 1937 effectively criminalized the plant at the federal level, and the American Medical Association, which had opposed the bill on the grounds that cannabis was a legitimate medicine, was steamrolled. When the Controlled Substances Act followed in 1970, cannabis was classified as Schedule I: no accepted medical use, high potential for abuse. Nixon's domestic policy advisor John Ehrlichman would later admit in a 1994 interview that the administration knew the science didn't support that classification, and did it anyway to criminalize the antiwar left and Black Americans.

The decades that followed were predictable. Underground use continued, research was stifled, and enforcement fell disproportionately on communities of color despite similar usage rates across racial groups. Honest information became scarce.

That started shifting in the 1990s. AIDS activists in San Francisco had spent years fighting for access to cannabis to manage the nausea and wasting associated with HIV treatment, and their organizing laid the groundwork for California's Proposition 215, which passed in 1996 as the first modern medical cannabis law. Colorado and Washington legalized recreational use in 2012. As of this writing, the majority of US states have legalized cannabis in some form, and the global picture is moving in the same direction. Legal markets have brought lab testing, regulated potency, and consumer labeling that simply didn't exist before.

They've also brought 40 mg gummies and 90% THC extracts, which is its own problem. The plant hasn't changed much. What's changed is what we do with it. That's worth understanding before you open the package.

How It Works

Every other substance in this book works by hijacking a receptor system that evolved for something else. MDMA commandeers your serotonin transporters. Ketamine blocks glutamate receptors. Cocaine jams your dopamine reuptake pumps. These molecules are essentially intruders, picking locks that weren't built for them.

Cannabis is different. Your brain has a dedicated receptor system for cannabinoids. It's called the endocannabinoid system, and it's been part of vertebrate biology for hundreds of millions of years. You produce your own cannabinoids naturally, the most studied of which is *anandamide*, named from the Sanskrit word for bliss. These endogenous cannabinoids act as a regulatory system, modulating mood, appetite, memory, pain response, and inflammation. They're part of the machinery your brain uses to maintain balance.

Unlike many of the other substances discussed in this book, the psychoactive effects of marijuana can't be easily summed up with a single molecule. Cannabis contains dozens of active cannabinoids (such as THC, CBD, CBG, and CBN), terpenes (e.g., limonene, β-myrcene, α-pinene, and linalool) and others, all of which contribute to the flavor and effects.

THC (*tetrahydrocannabinol*) is the primary psychoactive compound in cannabis, and the one most people know. It fits into the same receptors that anandamide uses: CB1 receptors, concentrated heavily in the brain and central nervous system, and CB2 receptors, found mostly in immune tissue throughout the body. When THC binds to CB1 receptors, it mimics and amplifies signals your endocannabinoid system would normally send, but with more intensity and for much longer than your brain's own molecules would sustain.

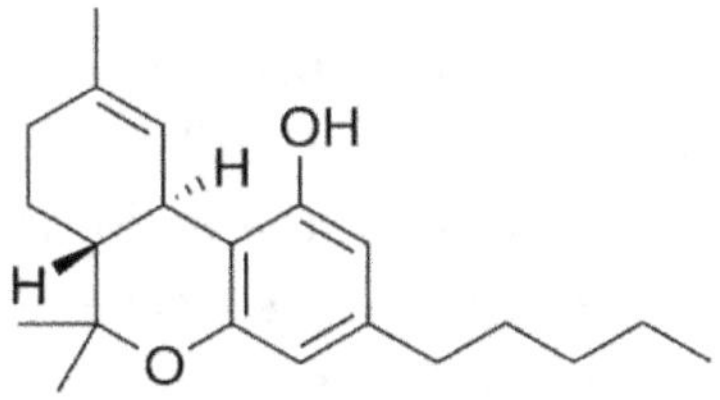

Tetrahydrocannabinol (Δ-9 THC)

That's the key distinction. Anandamide is broken down quickly. THC isn't. It lingers, keeps the receptor activated, and produces effects that outlast anything your natural system was designed to handle for long.

What you feel depends on where in the brain those CB1 receptors are firing. In the prefrontal cortex, THC disrupts working memory and warps time perception, which is why a twenty-minute conversation can feel like an hour and why you forget what you were saying mid-sentence. In the hippocampus, it interferes with forming new memories. In the amygdala, it modulates fear and anxiety, and this cuts both ways: low doses can quiet anxiety, while higher doses can amplify it dramatically.

THC also triggers dopamine release in the brain's reward circuits, which is where the pleasure comes from, and where the potential for habitual use lives.

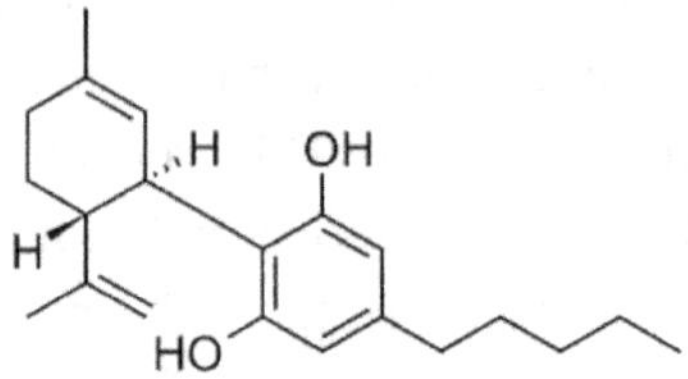

Cannabidiol (CBD)

Cannabidiol, better known as CBD, is a more complicated story. It doesn't bind directly to CB1 or CB2 receptors the way THC does. Instead, it interacts with the endocannabinoid system indirectly and has effects across several other receptor systems. The evidence suggests CBD can modulate anxiety, inflammation, and seizure activity, and there's decent reason to believe it can blunt some of THC's more uncomfortable effects, particularly anxiety and paranoia.

Modern cannabis products are not what was available ten years ago. Average THC content in legal market flower has climbed steadily, and concentrates can reach 70 to 90% THC. Cannabis in the 1970s typically ran between 1 and 3%. Today's legal market regularly offers products at 25 to 30%, and that's before you get to dabs and distillates.

This matters because the endocannabinoid system is dose-sensitive in ways that aren't intuitive. The relaxing, anxiolytic effects people associate with cannabis tend to occur at lower doses. At higher doses, the same system that was producing calm can tip into anxiety, paranoia, and in some cases, acute psychosis. That's not a personality flaw. It's a pharmacological threshold, and it's much easier to cross than it used to be. Know what you're working with before you start.

Indica, Sativa, and What the Label Won't Tell You

Walk into any dispensary and the first question you'll get is some version of "indica or sativa?" The shorthand is everywhere. Indica for the body, sativa for the head. Indica to sink into the couch, sativa to clean your kitchen at two in the afternoon. There's even a mnemonic for it: "indica" sounds like "in da couch." It's tidy, it's memorable, and as a guide to how a given product will actually make you feel, it's mostly wrong.

In the eighteenth century, botanists split the plant based on how it looked and where it grew. Cannabis sativa was the tall, lanky variety from equatorial regions, with narrow leaves and a long flowering cycle. Cannabis indica, described a few decades later, was the short, dense, broad-leaved type, associated with the Indian subcontinent and the harsh mountains of central Asia. Those descriptions are real, and they're still useful to a farmer deciding what to grow. They describe the architecture of the plant. They do not describe what happens in your brain when you smoke it.

The trouble is that almost nothing on a dispensary shelf is purely one or the other anymore. Decades of crossbreeding have blended the genetics so thoroughly that the labels on the jar are closer to marketing than to botany. When researchers have actually sequenced commercial strains, the genetic line between what's sold as indica and what's sold as sativa turns out to be blurry at best. Two jars carrying the same strain name, from two different growers, can be more different from each other than a random indica and sativa pulled off the same shelf.

Two things actually predict how a product will hit you. The first is the cannabinoid profile, mainly the ratio of THC to CBD, which we covered earlier. That ratio does more to shape your experience than the indica-sativa tag ever will. The second is terpenes, the aromatic compounds that give each variety its smell and, increasingly, its reputation.

Terpenes do more than make your weed smell like pine or citrus or diesel. Myrcene, the most common terpene in cannabis, is the one most associated with that heavy, sedative, sink-into-the-cushions feeling. It also happens to show up in high concentrations in a lot of what gets labeled indica, which may be where some of the folk wisdom came from in the first place. Limonene, bright and citrusy, is linked to mood elevation. Pinene, which smells exactly like it sounds, may push back against some of the short-term memory fog. Linalool, the lavender note, leans relaxing. This is what people call the "entourage effect," the idea that cannabinoids and terpenes work together to produce something none of them create alone.

None of this makes the indica-sativa question worthless. If a particular product labeled indica has reliably put you to sleep, that's real information about that product, and you should use it. Your own track record beats any label. What you shouldn't do is assume a different indica from a different source will do the same thing, or that reaching for a sativa guarantees you a clear, productive afternoon instead of a slow slide into anxiety.

Ignore the two-word binary as a promise. Read the actual numbers: THC percentage, CBD content, and the terpene breakdown if the product lists it, which more of them are starting to do. Pay attention to how specific products land for you and keep a mental note. That record, built from your own experience, will steer you better than the word stamped on the lid ever could.

The Edibles Problem

Here is the sentence that has ruined more evenings than almost anything else in recreational drug use: *"I don't think it's working."*

Edibles are the most misused delivery method in cannabis, not because they're inherently more dangerous than smoking, but because the gap between taking them and feeling them is long enough for people to make decisions they'll regret.

When you smoke or vaporize cannabis, THC enters your bloodstream through your lungs and reaches your brain within minutes. The onset is fast, the peak is relatively clear, and the whole experience is easier to calibrate.

Edibles work completely differently. The THC has to survive your digestive system before it can do anything. It passes through your stomach, gets absorbed in the small intestine, and then travels to your liver before it ever reaches your bloodstream. That process takes between thirty minutes and two hours depending on your metabolism, what you've eaten, and your individual biology. Some people report nothing for three hours.

That's not the only thing the liver does. During that first pass, THC gets converted into *11-hydroxy-THC*. This metabolite is more potent than THC itself, crosses the blood-brain barrier more efficiently, and produces a longer, heavier, more body-centered experience than inhaled cannabis. It's not just a stronger version of the same high. It's qualitatively different, and if you aren't expecting it, it can feel completely out of proportion to what you thought you took.

The combination of delayed onset and increased potency is where most edibles disasters come from. Someone takes a 10 mg gummy. Forty-five minutes pass. Nothing. They take another one. Sometimes two. Then the first dose arrives and the second is already on its way, and they are now significantly higher than intended with no way to reverse it.

This happens constantly, to seasoned cannabis users and first-timers alike. The solution is simple but requires patience that is hard to maintain when everyone around you seems fine.

Remember: you can always take *more*, but you can't take *less*. Start with 5 mg or less if you're new to edibles or trying a new product. Wait a full two hours before considering anything additional. If you're at a festival, wait longer. Heat, dehydration, and physical exertion all affect how your body processes THC. What felt manageable at home can feel overwhelming in a crowded field in the sun.

If you do find yourself more elevated than intended: you are not in physical danger. No one has died from THC overconsumption. It will pass. The timeline is uncomfortable but finite, typically four to eight hours for edibles, sometimes longer with high doses. Find somewhere quiet, drink water, eat something if you can manage it, and lie down if you need to. Put on a movie or a TV show you know well. The distraction of something comfortable can often work wonders. CBD, if you have it, may help take the edge off. Calm company helps more than almost anything else.

One more thing: tolerance plays a significant role with edibles in ways that aren't always obvious. A daily smoker trying edibles for the first time may find that their tolerance offers less protection than expected, because the metabolic pathway is different enough that smoking tolerance doesn't fully transfer. Start low anyway.

The edibles problem is entirely solvable. It just requires treating a gummy like the pharmacological object it is, rather than a snack.

Delta-8 and the Synthetic Gray Zone

If you've been in a gas station, a smoke shop, or a hemp store in the last few years, you've seen it. Delta-8 THC, sold legally in most states, marketed as a milder alternative to regular cannabis, often right next to the beef jerky and the energy drinks.

Natural cannabis contains *delta-9 THC*, the compound responsible for the high you're familiar with. *Delta-8 THC* is a closely related isomer, meaning the molecules have the same atoms arranged slightly differently. Delta-8 occurs naturally in cannabis, but only in trace amounts. Not nearly enough to be commercially useful.

The delta-8 in those gummies and vape carts isn't extracted from cannabis. It's synthesized, almost always from CBD derived from hemp, through a chemical conversion process involving acids, solvents, and reagents. The end product is technically delta-8 THC, but the process is unregulated, poorly standardized, and leaves room for a lot of things to go wrong.

Independent lab testing of commercially available delta-8 products has found residual solvents, heavy metals, and reaction byproducts that have no business being in something you inhale or eat. Because delta-8 exists in a legal gray zone rather than a regulated cannabis market, there are no consistent requirements for testing, labeling, or purity. The certificate of analysis on the website, if there is one, may not reflect what's actually in the product you're holding.

The effects of delta-8 are real. It does bind to CB1 receptors and produce a psychoactive effect, generally described as less intense and less anxiety-provoking than delta-9. That characterization is broadly accurate at moderate doses, which also means people tend to use more of it. The dose-response curve is not as forgiving as the marketing implies. Adverse events including anxiety, paranoia, hallucinations, and loss of consciousness have been documented by poison control centers in states where delta-8 is widely sold.

Delta-8 products are heavily marketed toward people who either can't access legal cannabis or are specifically looking for something that won't show up on a drug test. (It does show up on a drug test. Standard panels don't distinguish between isomers.)

Despite the advertising, delta-8 is not a safe or regulated alternative to cannabis. It's a chemically derived product sold in an unregulated market with inconsistent quality and real potential for contamination. If you're in a legal cannabis state, tested products are a much safer bet than unregulated ones. That gas station gummy is a worse bet than it looks.

What Can Actually Go Wrong

Cannabis has a forgiving safety profile, but that doesn't mean that things can't go wrong. Here's what to do if it does.

Anxiety and Paranoia

This is the most common adverse effect of cannabis, and it's almost entirely dose-dependent. The same endocannabinoid system that produces calm and euphoria at lower doses can tip into anxiety, racing thoughts, and paranoia when pushed past a certain threshold. That threshold varies significantly between people and is affected by tolerance, set and setting, the specific product, and whether anything else is in your system.

High-THC products make this more likely. So does being in an unfamiliar or overstimulating environment. So does using cannabis to manage pre-existing anxiety, which works until it doesn't, and then it *really* doesn't.

If you or someone with you hits that wall, the approach is similar to navigating a bad psychedelic moment, because at high doses the experiences have more in common than most people expect. Move to a quieter space. Sit or lie down. Breathe slowly. Put on a familiar movie or TV show. Remind yourself or them that this is temporary and pharmacological, not a permanent state or a reflection of reality. Don't fight it. Let it move through.

CBD is worth having specifically for this scenario. There's reasonable support for the idea that CBD can moderate THC-induced anxiety by interacting with some of the same receptor pathways. A CBD tincture or capsule taken at the onset of a difficult experience may help flatten the spike. It won't end the trip, but it can take the sharpest edges off.

One thing not to do: don't give someone in a cannabis anxiety spiral more cannabis, even if they ask for it. The instinct to chase relief with more of the same substance is a trap. It rarely helps and often makes things considerably worse.

Cannabinoid Hyperemesis Syndrome

CHS is one of the stranger things in harm reduction, and most people who have it don't know what it is until someone tells them.

Cannabinoid Hyperemesis Syndrome, characterized by cycles of severe nausea, vomiting, and abdominal pain that can last for days, develops in some long-term, heavy cannabis users. What makes it strange is that cannabis is widely understood to reduce nausea. CHS is essentially the opposite effect emerging after extended heavy use, and the mechanism isn't fully understood.

The most distinctive feature of CHS is that hot showers and baths provide dramatic, temporary relief. Nobody knows exactly why. The leading theory involves temperature-sensitive receptors in the skin that interact with the endocannabinoid system, but the research is still catching up. What's documented is that people with CHS will sometimes spend hours cycling in and out of hot showers because it's the only thing that helps.

The only effective treatment for CHS is stopping cannabis use. Symptoms typically resolve within days to weeks of cessation. If you recognize this pattern in yourself or someone you know, including the hot shower compulsion, that's the diagnosis. Get it checked out and take the cessation recommendation seriously.

CHS is more common than most people realize, partly because it's underdiagnosed and partly because the cannabis-nausea connection is so counterintuitive that people don't connect the dots for a long time.

Synthetic Cannabinoids

This section exists because synthetic cannabinoids are not cannabis, do not behave like cannabis, and have killed people who thought they were using cannabis.

Synthetic cannabinoids, sold under names like K2, Spice, and a rotating cast of brand names designed to stay ahead of scheduling laws, are lab-created compounds that bind to CB1 receptors with much higher affinity than THC. That sounds like it might just mean a stronger high. In practice it means an unpredictable, potentially dangerous time with a much narrower margin for error.

Effects can include extreme agitation, psychosis, seizures, dangerously elevated heart rate, kidney failure, and death. These are documented outcomes at doses that someone expecting a cannabis experience would consider normal.

Synthetic cannabinoids are most commonly encountered in products sold as herbal blends or legal highs, but they also show up in festival settings, sometimes sold as or mixed with actual cannabis. They are not detectable by smell or appearance, and standard reagent testing kits don't reliably identify them.

The harm reduction here is limited by the absence of a reliable field test. What you can do is know your source. Buy from licensed dispensaries in legal markets where products are tested. If someone is having a reaction that seems wildly out of proportion to what they took (respiratory distress, seizure, extreme confusion, or loss of consciousness), that is a medical emergency. Call for help immediately.

The Psychedelic Amplification Risk

At low doses, cannabis can soften the edges of a psychedelic trip, adding warmth or relaxation in the latter stages when the intensity is fading. That effect is real.

At higher doses, or if timed poorly, cannabis can dramatically intensify a psychedelic journey in ways that are difficult to manage. This is especially true near the peak, when your brain's filtering systems are already disrupted. Adding THC in that window can trigger a sudden escalation that feels completely disproportionate and can tip a manageable time into a difficult one very quickly.

The particular risk is with edibles and psychedelics together. Someone who takes an edible early in a psychedelic session, feels nothing for an hour, and forgets about it is in for a surprise when both peaks arrive simultaneously. This sends more people to festival medical tents than almost anything else on this list.

If you're going to combine cannabis with a psychedelic, use it sparingly, use a known dose, and wait until you're well past the peak. If you're new to either substance, don't combine them until you know how each affects you individually.

Safer Use

Most of what makes cannabis safer comes down to three things: knowing your dose, knowing your product, and knowing yourself.

Start low, go slow. This applies to every delivery method, but especially edibles. With flower or vaporizers, that means a single inhalation and a few minutes before deciding you need more. With edibles, it means starting at 5 mg or below and waiting a full two hours. The phrase sounds like a bumper sticker because it works.

Know your product. In a legal market, read the label. THC percentage matters, and so does the ratio of THC to CBD. A product with meaningful CBD content will generally be more forgiving than one that's pure high-potency THC. If you're buying from a dispensary, ask the staff. Outside legal markets, know your source and be skeptical of anything that looks, smells, or behaves differently than expected.

Know your tolerance... but don't trust it completely. Tolerance to cannabis builds relatively quickly with regular use, but it's delivery-method specific in ways that catch people off guard. A daily smoker trying edibles for the first time may find their tolerance offers less protection than expected. If you've taken a break of even a few weeks, your baseline has dropped. Adjust accordingly.

Set and setting also matter more than people expect. Cannabis is not a neutral substance when it comes to environment and mindset. High-THC products in particular are sensitive to context. An anxious headspace, an overstimulating environment, or a social situation where you feel uncomfortable can push an experience toward paranoia faster than the same dose in a calm, familiar setting. Heat, dehydration, sleep deprivation, and sensory overload lower your threshold. What you handle fine at home may hit differently at hour fourteen of a festival day. Drink water. Eat something before you use cannabis. Find shade if you're outside.

Have CBD on hand. If you use cannabis regularly, especially in settings where you might push the dose or encounter something stronger than expected, keeping a CBD tincture or capsules nearby is worth the small investment. Twenty-five to fifty milligrams taken at the onset of an uncomfortable experience is a reasonable starting point.

Avoid combining with other substances until you know both well. Cannabis interacts with alcohol, increasing intoxication and nausea more than either does alone. It interacts with psychedelics in ways that are hard to predict. It can amplify stimulant anxiety and extend dissociative experiences. If you're using cannabis alongside anything else, use less of both than you think you need.

If someone is having a hard time: move to a quieter space, sit with them, speak calmly. Remind them what they took, that it's temporary, and that they're physically safe. Don't leave them alone if they're significantly distressed. Don't give them more cannabis. Offer water, food if they want it, and CBD if you have it. Most cannabis overconsumption incidents resolve on their own within a few hours with nothing more than calm company and a comfortable place to be.

If someone is unresponsive, seizing, or showing signs of extreme physical distress rather than psychological discomfort, that warrants medical attention. As noted in the synthetic cannabinoids section, reactions that seem dramatically out of proportion to what someone took should be taken seriously.

Supplements and Recovery

Cannabis doesn't deplete your neurotransmitters the way MDMA does or torch your B-12 like nitrous oxide. The supplement considerations here are more modest, and the research specifically targeting cannabis users is thinner than what exists for other substances in this book. What follows is grounded in plausible mechanisms and solid general evidence, with that caveat noted where it applies.

Before (if you're anxiety-prone or trying something strong)

- **CBD (25-50 mg)** taken 30 to 60 minutes before use may help moderate THC-induced anxiety. The evidence is credible enough to recommend. Look for a broad or full-spectrum product from a reputable source with third-party testing.
- **Magnesium glycinate (200 mg)** is worth taking if you know you're sensitive to anxiety or planning to use in a stimulating environment. Same reasoning as the MDMA chapter: a general anti-anxiety supplement that supports the nervous system without sedating you.

During (if things go sideways)

This is where the folk remedies live, and two of them are interesting enough to include despite thin formal evidence.

- **Black pepper** is the most documented. Chewing a few whole peppercorns or sniffing freshly ground pepper during a cannabis anxiety spiral has substantial anecdotal support, and there's a plausible mechanism behind it. Black pepper contains *beta-caryophyllene,* a terpene that binds to CB2 receptors and may help modulate the anxiety response. Keep some in your kit.

- **Lemon peel** works along similar lines. *Limonene*, the terpene responsible for that sharp citrus smell, may have anxiolytic properties through serotonin and dopamine pathways. Smelling or chewing fresh lemon peel is the standard application.
- **CBD (25-50 mg)** if you have it. Start low and work up. Look for sublingual drops, if you can find them. They'll kick in the fastest.

After (for regular or heavy users)

- **Omega-3 fatty acids** are the most evidence-backed recommendation here. Your endocannabinoid system uses omega-3-derived fatty acids as precursors to produce your natural endocannabinoids. Heavy cannabis use downregulates CB1 receptors over time, and omega-3 supplementation supports receptor health and endocannabinoid tone during tolerance breaks or cessation. If you use cannabis regularly and aren't already taking omega-3s, this is the supplement most likely to make a noticeable difference. Standard dosing is 1-2 grams of combined EPA and DHA daily.
- **Citicoline (CDP-choline, 300-600 mg)**: Heavy or long-term use is associated with downregulation of CB1 receptors and some reduction in dopamine signaling in the prefrontal cortex. Citicoline supports acetylcholine production, has well-documented neuroprotective properties, and there's preliminary evidence it supports dopamine receptor density. For regular users experiencing cognitive fog, reduced motivation, or memory issues, which are the most commonly reported cognitive side effects of chronic heavy use, citicoline is a reasonable addition to your stack.

- **Melatonin (1-5 mg)** for sleep. Cannabis suppresses REM sleep with regular use, and when heavy users reduce or stop, rebound insomnia and vivid dreams are common. Melatonin helps bridge that adjustment period.
- **Magnesium glycinate (200 mg)** before bed, for sleep quality and nervous system support during any tolerance break or period of reduced use.

A Note About CHS

If you're experiencing Cannabinoid Hyperemesis Syndrome, the primary treatment is stopping your use. There are no supplements that treat it. Capsaicin cream applied topically to the abdomen has shown some relief in clinical case reports, likely through the same receptor pathway that explains why hot showers help. It's a more practical stopgap than standing in the shower for three hours. Use it alongside, not instead of, addressing the underlying issue. CHS is also a useful reminder: cannabis is a *real drug*. If your use has gotten away from you, help is at the back of the book.

What Cannabis Offers

Cannabis is the most democratized psychoactive substance on earth. It's legal in more places than it's ever been, more potent than it's ever been, and more casually used than it probably should be.

At its best, cannabis is genuinely useful. It can quiet anxiety, soften physical pain, deepen music, open conversation, and for some people, provide a reliable on-ramp to the kind of present-moment awareness that meditation practitioners spend years cultivating. In therapeutic contexts, it's showing real promise for PTSD, chronic pain, and end-of-life care. These aren't small things.

It can also become a habit that quietly insulates you from your own life. Not dramatically, not in the way that makes for compelling intervention television, but gradually. The motivation that softens. The emotions that stop moving through and start backing up instead. The mornings that require it before they feel manageable. Cannabis dependency doesn't announce itself the way other dependencies do, which is part of what makes it easy to overlook until it isn't.

None of that makes it dangerous in the way fentanyl is dangerous, or demanding in the way MDMA is demanding. It just makes it worth paying attention to, which is the same thing this book has asked of you in every chapter.

The substances that are easiest to dismiss as harmless are often the ones that benefit most from a little honest attention. Cannabis has been with humans for twelve thousand years. It's been used reverently and recklessly, medicinally and recreationally, sacredly and stupidly. It will outlast every law written about it and every moral panic directed at it. It doesn't need your reverence. It just deserves your respect.

Start low. Know your source. Keep some black peppercorns around. And if the edible hasn't kicked in yet, put the second one down and wait.

Nitrous Oxide – "Whippits"

"It's always tea time."

A Brief History

Nitrous oxide came out of Enlightenment chemistry and has been tangled in science, medicine, performance, and play ever since.

It began in 1772, when British chemist Joseph Priestley first synthesized the gas by heating ammonium nitrate. He called it "phlogisticated nitrous air," a name only a pre-modern chemist could love. It wasn't until 1799 that things got interesting. That year, a young chemist named Humphry Davy began experimenting with it, taking increasingly large doses and inviting his friends to do the same. He noticed that the gas produced giddiness, laughter, pain relief, and vivid alterations in thought. Davy, half-laughing and half-serious, dubbed it "laughing gas" and proposed its use as a surgical anesthetic.

His suggestion was ahead of its time. Medical science wasn't ready to take the idea seriously, but performers were. By the early 1800s, nitrous had become part of traveling entertainment. Audiences would watch as volunteers stumbled on stage, slurring their speech and laughing uncontrollably under the influence of the gas. What was meant as a spectacle occasionally revealed something deeper; when participants injured themselves during these demonstrations, they often felt no pain until the gas wore off.

It was this very phenomenon that changed history. In 1844, American dentist Horace Wells attended one of these shows and realized what he was seeing: analgesia. He arranged to have a tooth extracted under the influence of nitrous oxide and felt nothing. Though his public demonstration would later falter, the idea had taken root. Within decades, nitrous oxide became a staple of dental and surgical anesthesia, sometimes alone, often in combination with other agents.

Even as its medical legitimacy grew, nitrous remained dual-natured. It moved fluidly between the operating room and the underground. In the 1960s and '70s, psychonauts began exploring its use alongside LSD or cannabis to enhance sensory distortion and emotional lability. By the 1990s, the rise of whipped cream canisters and cartridges ("whippits") made it more accessible than ever. Cheap, discreet, and fast-acting, nitrous became a mainstay at raves, festivals, and basement parties.

Even today, nitrous is legally available, widely used, and vastly underestimated.

The Experience

Thirty seconds. That's the window. It's shorter than most people expect, and stranger than most people are prepared for.

The onset is almost instantaneous. Within one or two breaths, a warm pressure spreads through the chest and rises into the head. Sound warps first: voices slow and distort into a low, cycling "wah-wah-wah," like a record player losing speed. (In Australia and New Zealand, they call nitrous chargers "*nangs*," after this onomatopoeic sound.) Your own thoughts start to echo. The room tilts. Everything becomes slightly too bright, slightly too funny, and slightly too significant.

Then time does something strange.

Those thirty seconds stretch. Not in an unpleasant way, but in the way that deep sleep stretches: you emerge from it feeling like more happened than the clock allows. Some people describe it as looping, a sense that the same moment is repeating, cycling through itself on a closed track. Others describe an inexplicable nostalgia, a feeling of returning to somewhere familiar that they can't name. Many people laugh, not because anything is funny, but because the release of ordinary cognition is physically relieving.

At higher doses, taken in close succession, the experience deepens into something more disorienting. Identity begins to blur. Language stops working. The sense of being a specific person in a specific place softens and, briefly, disappears.

The philosopher William James wrote about this territory in *The Varieties of Religious Experience.* He had experimented with nitrous oxide himself and was struck by how powerfully the gas produced a sense of mystical insight and cosmic unity: the feeling that everything was reconciled, that opposites resolved into each other, that the whole thing finally made sense. He called it the "anesthetic revelation." What troubled him was that the moment consciousness returned, the insight evaporated. What remained was only the absolute certainty that it had been there, combined with the total inability to say what it was. James found this philosophically significant. The experience raised the question of what kind of knowing the ordinary mind is capable of, and what it might be filtering out the rest of the time.

Most people who use nitrous at parties are probably not thinking about William James, but the strangeness he described is exactly what everyone is chasing, even if they're calling it something else.

What It Does and How

Nitrous Oxide

Nitrous oxide works fast. Really fast. You inhale, and within seconds your ears are ringing, your vision is swimming, and your body feels like it just left the building, but what's actually happening inside your brain?

Your brain runs on electrical signals. As we've seen, these signals jump between brain cells, using chemicals known as neurotransmitters. Some of these chemicals tell the next cell to fire, like shouting "Go," while others act as brakes and say, "Hold on." One of the main chemicals that speeds things up is called glutamate. Glutamate helps keep your brain alert, focused, and responsive to the world around you.

Nitrous interferes with that system. It blocks a specific type of glutamate receptor called *NMDA*. Think of this receptor as a door that allows signals to pass between brain cells. When nitrous enters the picture, it wedges that door shut. Suddenly, the flow of information stops getting through as smoothly. Your senses all start to blur and stutter. Everything starts to feel strange, disconnected, or delayed.

This is what causes dissociation. You might feel like you're floating outside your body, or like your thoughts have slowed to a crawl. Some people experience looping sounds, visual distortions, or a sense that time has frozen. You're not hallucinating in the traditional sense, just processing the world in a way your brain doesn't normally allow.

That's only part of the story. Nitrous also stimulates the release of dopamine, which is your brain's reward chemical. This creates a wave of euphoria and explains why people often reach for a second balloon before the first one has even worn off. It also activates parts of the opioid system, giving a feeling of warmth, comfort, and pain relief. That mix of mental quiet and emotional pleasure is what makes nitrous feel so appealing and so easy to overuse.

Safer Use

Respecting nitrous means understanding both the acute dangers and the long-term consequences. Unlike many substances, nitrous oxide doesn't feel threatening. It doesn't usually cause nausea, anxiety, or dysphoria. It doesn't last long enough to derail your night, but that disarming quality can make its risks more insidious. The first rule of nitrous is simple: don't underestimate it just because it's fun. To understand why nitrous oxide can sneak up on you, we have to start with B-12.

The B-12 Problem

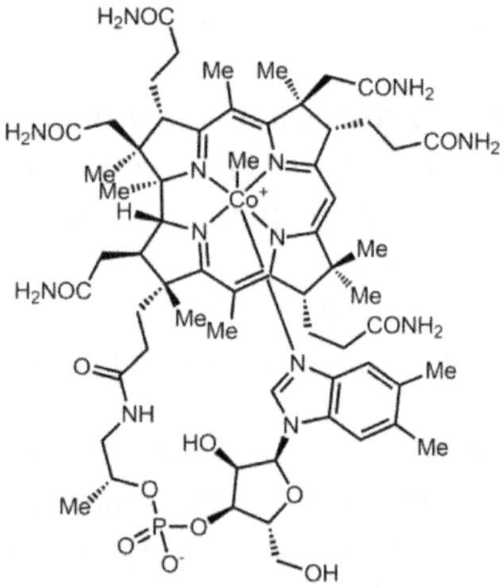

Vitamin B-12 has a charged cobalt (Co^+) atom in the center

Vitamin B-12 is one of the most complex and critical vitamins that your body requires. It has a charged cobalt atom at its core, and nitrous oxide targets that atom with unsettling precision. When nitrous oxidizes the cobalt, it flips B-12 into an inactive form that your body can no longer use, no matter how much of it is circulating. This matters because B-12 is one of the quiet custodians of your nervous system. It fuels the production of *myelin*, the protective insulation wrapped around your nerves. Think of myelin as the rubber coating on electrical wiring. Strip that coating away and the signal weakens, stutters, or misfires. The same thing happens in your nerves. Without enough active B-12, the body struggles to maintain that insulation, and the earliest signs show up as tingling, numbness, or a subtle wobble in your gait. Left unchecked, the damage can deepen into real neurological injury because the nerves themselves begin to fray when the sheath around them breaks down.

To prevent this, supplement aggressively if you're using nitrous with any regularity, and check the form of B-12 you're taking. You want *methylcobalamin*, not *cyanocobalamin*, ideally in sublingual or injectable form for better absorption (since it bypasses digestive breakdown). Pairing B-12 with *L-methionine* (an amino acid that supports the same methylation pathways) may offer additional neuroprotection.

Early Warning Signs

The neurological damage from nitrous accumulates slowly, rather than arriving all at once. By the time most people recognize the problem, it's been building for a while. Knowing what to watch for early is the difference between catching it and suffering it.

The first signs are usually in the extremities. Tingling or numbness in the hands and feet, especially in the fingertips, is one of the earliest indicators that B-12 depletion is affecting nerve function. Some people notice it around the mouth as well. It may come and go at first, and it's easy to dismiss as poor circulation or fatigue. Don't dismiss it.

Other early signs include unusual fatigue that doesn't resolve with sleep, difficulty with balance or fine motor tasks, brain fog or difficulty concentrating, and mood changes including increased irritability or low-grade depression. These are less specific, but in the context of regular nitrous use, they're worth taking seriously.

The question everyone asks is: how often is too often? There's no clean threshold, but here's a reasonable framework. A single recreational session for someone with adequate B-12 stores is low-risk. Weekly use, or heavy single sessions involving many dozens of cartridges, is where meaningful depletion begins. If you are vegan, vegetarian, or elderly, your baseline B-12 stores may already be lower than average, which makes you more vulnerable at lower frequencies of use.

If you use nitrous more than a few times a month, get bloodwork done. Ask your doctor to check B-12 levels, homocysteine, and MMA (methylmalonic acid). The last two are more sensitive markers of functional B-12 deficiency than serum B-12 alone. Catching a deficiency early is straightforward to treat. Catching it late is a different conversation.

If any neurological symptoms appear, including tingling, numbness, balance issues, or cognitive changes, stop using nitrous and see a doctor before you use it again.

Nutritional Support

Beyond inactivating B-12 in isolation, nitrous disrupts a broader biochemical process called the methylation cycle, which B-12 sits at the center of. Understanding that cycle a little helps you support it intelligently, because B-12 alone isn't always enough.

The methylation cycle is how your body produces and recycles the molecules it needs for DNA repair, neurotransmitter synthesis, and nerve function. B-12 is the key, but folate, B-6, and B-2 are the supporting cast. When nitrous knocks out B-12, it throws the whole cycle off balance. Supplementing the full pathway gives your body more to work with.

Here's what can help:

Methylcobalamin B-12 (1000-2000 mcg, sublingual): The sublingual form bypasses the digestive system and absorbs directly into the bloodstream. This matters because oral B-12 depends on a protein called intrinsic factor, which some people produce in lower amounts. Take it before and after sessions. If you use nitrous regularly, take it daily. B-12 is very difficult to overdose on. In this case, less is not more.

Folate as methylfolate (400-800 mcg): Not folic acid, the synthetic version. Look for methylfolate or 5-MTHF on the label. This is the bioavailable form your body can most easily use. Folate works directly alongside B-12 in the methylation cycle and supports the same nerve repair pathways.

Vitamin B-6 as P-5-P (25-50 mg): Pyridoxal-5-phosphate is the active form of B-6. It's a cofactor for neurotransmitter synthesis and works alongside B-12 and folate in the methylation cycle. It may also help with mood stability in the days following heavy use.

Riboflavin/Vitamin B-2 (100 mg): The most overlooked member of this stack. B-2 is essential for converting folate into its active form. Without enough B-2, your folate supplementation doesn't work as well. It's cheap, safe, and rarely causes problems.

L-Methionine (500 mg): An amino acid that supports the methylation pathway downstream of B-12. It helps your body continue producing SAM-e, the molecule the methylation cycle ultimately generates. Take it with or after sessions rather than before.

You don't need to take all of these every time you use nitrous. But if you're using with any regularity, a daily B-complex that covers methylcobalamin, methylfolate, B-6 in P-5-P form, and B-2 is a reasonable baseline. Think of it as keeping the tank topped off rather than scrambling to refill it after the fact.

Additional Harm Reduction

Beyond biochemistry, hypoxia (lack of oxygen) is the most immediate concern. Nitrous displaces oxygen in the lungs. Taking repeated hits from a balloon without breathing fresh air can lead to dizziness, tunnel vision, blackouts, or even unconsciousness. Blue lips or nails are a sign of hypoxia, and mean that you're not getting enough oxygen to your brain. Always inhale, then take a few breaths of normal air before the next one. Never strap a mask to your face or inhale directly from a whipped cream dispenser or medical tank. The gas is ***very*** cold when it comes out of the tank (as low as -40°C/F!), and can burn or instantly freeze your skin or airway.

Many injuries from nitrous don't come from the gas itself, but from people collapsing mid-inhale, hitting their head, or dropping heavy objects as they fall. Even a few seconds of unconsciousness is long enough to break a tooth, split a lip, or crack a skull. If you use nitrous, do it seated, lying down, or somewhere stable where you cannot fall. Avoid standing, dancing, or leaning over hard surfaces. Never inhale while driving, standing on concrete, or sitting near sharp corners. That sudden blackout can come faster than you expect, and gravity always wins. Safety here is simple: stay low, stay still, and let your body catch up before you take another breath.

Never use nitrous oxide in an enclosed area. The gas itself is not poisonous, but it pushes oxygen out of the air you need to breathe. Inside a car, tent, closet, or under a blanket, it can quietly starve your brain of oxygen before you even notice. Passing out from lack of air is not a funny story; it's how people die. The lightheaded feeling that many chase is the first sign that oxygen is running low. If you stay in that environment, the next step is unconsciousness, and then suffocation. Always use in a space with plenty of fresh air. Crack a window, step outside, and never do it alone. Nitrous can make you feel far away from your body, but it cannot protect it. Oxygen is not optional.

If you're using whipped cream cartridges ("whippits"), always use a balloon. The gas exits the cartridge under high pressure and at extremely low temperature. Inhaling it directly is dangerous and unnecessary. If you're using a whipped cream cartridge cracker, allow the gas to fill a balloon and sit for a moment. This allows the cold to dissipate and the oils or lubricants (often present in food-grade cartridges) to settle. Never inhale the first hit from a new cartridge if you can avoid it; that's where the impurities concentrate.

More recently, large-format cartridges have become common at parties. The 580 g and 640 g tanks hold the rough equivalent of 100 or more standard whippits. Where a box of chargers once created a natural pause between sessions, a single large cartridge removes that friction entirely. The B-12 math changes significantly when you can move through that volume in a single night. If this is the format you're working with, treat it as a higher-risk activity. Set a limit before you start. Stick to it.

Hygiene matters. Reuse balloons only if they're clean and intact. Old latex can harbor bacteria or break mid-hit, posing a choking hazard. Keep your equipment dry and away from heat sources. Avoid using in cars, closets, or enclosed spaces where oxygen displacement is a bigger risk.

Then there's pattern awareness: Nitrous is unusual among psychoactives in that it invites compulsive re-dosing, not because it's intensely pleasurable in a classic sense, but because the effects are fleeting. Each balloon lasts maybe 30 seconds. The temptation is to keep chasing that floaty, giggly dissociation by stringing balloons together one after another. And then again. And again.

This is where use becomes risky. Binge sessions, especially those lasting hours or involving dozens of cartridges, create the conditions for neurotoxicity. If you find yourself going through a box every night, hiding use from others, or choosing nitrous over more fulfilling activities, it's time to take a step back. Just because a drug is legal and cheap doesn't mean it's harmless.

Finally, remember that nitrous has a cultural invisibility. It's rarely taken seriously in conversations about drug safety. That can make it harder to recognize when you or someone you care about has crossed into dangerous territory. If someone's walking with a limp, slurring their speech days after a party, or complaining of numbness, don't dismiss it as burnout. Get bloodwork done. Look at B-12 levels. Catch it early.

Nitrous can be part of a joyful, celebratory, or even introspective practice, but only if you treat it as something real, because **it is.** It affects your brain, your nerves, your lungs, your judgment. It deserves the same level of respect and care we bring to substances that come with more obvious reputations.

Give yourself some space between your sessions. Supplement before and after. Keep oxygen in your rotation. And above all, stay aware of your patterns. Insight doesn't just come from what you take; it comes from how you take it.

What Nitrous Offers

Nitrous is the easiest substance in this book to underestimate. It's legal. It's at every festival. It's over in thirty seconds. People write it off as the drug you use while waiting for the other drugs to kick in.

That framing misses something real.

At social doses, nitrous is light and joyful. It's communal. People laugh. Tension dissolves. It has a long history as a substance of celebration precisely because it delivers a brief, clean window of release without the weight that other substances carry.

At higher doses, taken in sustained sessions, the territory changes. The dissociation deepens into something more genuinely liminal. The experience William James described, that sense of deep reconciliation and unity that evaporates completely on return to ordinary consciousness, is not a trivial phenomenon. It shows up consistently, across cultures, in people who aren't expecting it and don't have a framework for it. The insight doesn't survive the return to sobriety. What survives is the certainty that it was there.

Whether that means anything is up to you. But it's worth taking seriously, in the same way any altered state that touches questions of identity and meaning deserves to be taken seriously.

The problem is that because nitrous is so short and so accessible, it almost never gets that seriousness. It becomes a tic. A nervous habit. A thing you do between conversations. The doses stack up, the B-12 depletes quietly, and by the time you notice, the issue has been building for months.

Use less, space your sessions, and supplement the methylation cycle before and after. Nitrous will always be there at the next party. The question worth asking isn't how many balloons you can fit into a night; it's whether you're getting anything out of them.

Cocaine

"Now, here, you see, it takes all the running you can do, to keep in the same place. If you want to get somewhere else, you must run at least twice as fast as that!"

Origins and Cultural Trajectory

Long before the white powder hit mirrored tables and bathroom stalls, coca was a leaf. To some, it was a gift from the mountains, sacred to those who lived in their shadow. In the high Andes of South America, the Quechua and Aymara peoples chewed coca leaves for millennia. These weren't recreational highs, but part of daily life. Coca staved off hunger and cold. It eased altitude sickness, brought stamina to long walks and hard labor. In ritual contexts, coca was a divine offering to the *Apus*, the mountain spirits, and a conduit to the ancestors.

Unlike refined cocaine, whole coca leaf delivers a gentle, sustained stimulation, similar to a cup of coffee. It was used communally, often shared in *ayllu* (kinship groups), and chewed with lime to enhance absorption. Spanish colonists, recognizing coca's utility, didn't outlaw it, but taxed it instead. Miners in the Potosí silver mines were required to chew coca for longer endurance in brutal, oxygen-poor tunnels. Colonial empires turned a sacred leaf into an instrument of extraction.

It wasn't until the mid-19th century that European scientists isolated the active alkaloid, *cocaine hydrochloride*. The purified compound was far more potent than the leaf. Its effects were immediate, euphoric, and deeply reinforcing. In the 1860s, German chemist Albert Niemann described its numbing and stimulant properties, and by the 1880s, medical and commercial interest exploded.

One of cocaine's earliest celebrity endorsements came from Sigmund Freud, who used it extensively and recommended it to colleagues as a cure for depression, indigestion, and even morphine addiction. He would later distance himself from the substance, having witnessed friends suffer breakdowns under its spell.

At the same time, John Pemberton, a wounded Civil War veteran and morphine addict, was developing a patent medicine in Georgia that combined coca leaf extract with caffeine-rich kola nut. He called it French Wine Coca, modeled after the popular European coca wines like Vin Mariani. When Prohibition laws arrived in Atlanta, he reformulated it, removing the alcohol but keeping the coca, and thus created the original formulation of Coca-Cola.

This early Coca-Cola did, in fact, contain active cocaine. The exact amount is debated, but the stimulant kick was real, and part of its appeal. It wasn't until 1904, under mounting public pressure and early narcotic regulations, that the company shifted to a "decocainized" coca leaf extract. Even today, the Coca-Cola Company continues to import coca leaves legally through a special arrangement with the Stepan Company, which extracts the flavor while transferring the active alkaloid to government-licensed pharma, where it is refined and used legally for nasal surgery.

As the 20th century progressed, cocaine moved from pharmacies to nightclubs. In the 1970s, it became the drug of choice in American disco culture, synonymous with flash, freedom, and an emerging culture of sensual self-expression. In the crowded floors of Studio 54, cocaine was as common as champagne. It was seen as sophisticated, not stigmatized, an elite drug for elite people. This was a stark contrast to its portrayal just a decade later.

In the 1980s, as Wall Street roared and wealth became spectacle, cocaine rebranded again, this time as a power drug. Traders, executives, and creatives took lines between deals, meetings, and studio takes. The high matched the pace of the era: fast, confident, relentless. Just beneath the gloss, the cracks began to show. Addiction rates soared. Paranoia and burnout became epidemic in professional circles.

Meanwhile, crack cocaine, a cheaper, smokable form, hit urban communities with devastating speed. The same molecule that graced boardrooms tore through families and neighborhoods, especially in Black and Latino communities. The government responded with punishment, not public health. Harsh sentencing laws treated crack cocaine as fundamentally more criminal than powder, despite pharmacological similarities.

That split carries through to today. Powder cocaine remains coded as high-status, while crack carries stigma and criminality, but both share the same root. Both descend from the sacred coca leaf. And both tell the story of how a molecule, once chewed in reverence, became a symbol of escape, ambition, and ruin.

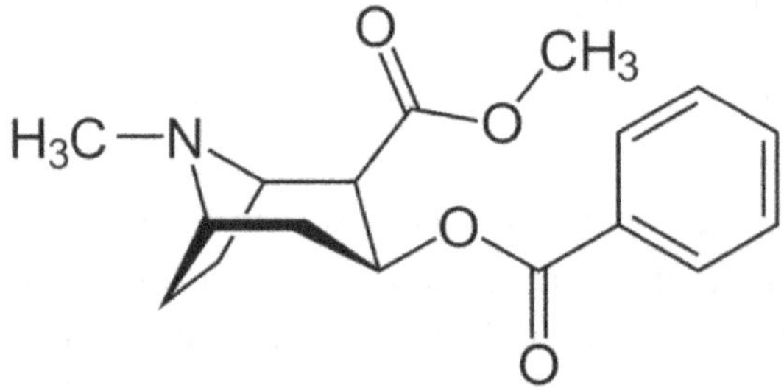

Cocaine

How It Works: Neurochemistry of Desire

Cocaine seizes the machinery that already creates desire. It takes the brain's reward circuit, built for subtle pulses of motivation and satisfaction, and pushes it far beyond its natural range.

Under normal conditions, dopamine is released into the space between brain cells, the synapse, where it carries its message. Then, as part of the reuptake process, it's quickly reabsorbed by the cell that released it. This cycle keeps your brain's reward system balanced and responsive.

Cocaine blocks that reuptake.

Instead of allowing dopamine to return and reset, cocaine traps it in the synapse. The signal stays stuck in the "on" position. Your brain keeps hearing that same message, again and again, louder and longer than it was meant to. You feel focused, energized, euphoric, and powerful. For a little while, everything seems sharper and easier.

Dopamine is only part of the story. Cocaine also affects norepinephrine, your body's internal alert system. It increases heart rate and blood pressure and puts you into a state of heightened arousal. This is the same system that activates when you're in danger. Cocaine flips that switch whether there's a threat or not. Even if you're sitting still, your body thinks it's preparing to sprint, fight, or escape. This is part of what gives cocaine its intensity and also what makes it hard on your heart.

While cocaine also raises levels of serotonin, its effect on serotonin is less dramatic than what it does to dopamine and norepinephrine. The dopamine flood is what defines the cocaine high.

The brain, however, doesn't like being pushed out of balance. If dopamine is left floating in the synapse too long, your brain starts to adapt. It dials back the sensitivity of dopamine receptors. It tries to restore order. The more often you use cocaine, the more your brain resists the signal. Over time, natural sources of pleasure (like food, music, sex, or connection) start to feel muted. Nothing quite compares to the chemical surge cocaine delivers.

This is how addiction begins. Cocaine trains your brain to need more of what it just had. And because the high wears off quickly, often in less than an hour, the crash can feel sudden and severe. That crash is chemical as much as emotional. Dopamine levels drop, and your brain struggles to recalibrate. You may feel anxious, irritable, hollow, or desperate; that is when the craving to redose kicks in.

Meanwhile, your cardiovascular system is under increasing strain. Cocaine tightens blood vessels and pushes your heart to work harder. Combined with elevated blood pressure and body temperature, this creates a real risk of heart attack or stroke. These effects can hit even young or otherwise healthy users, especially when doses pile up or cocaine is mixed with other substances.

Cocaine accelerates whatever is already moving inside you. It pushes your existing emotions into overdrive, whether that is confidence, restlessness, insecurity, or fear. This is why it can feel revelatory to one person and destabilizing to another. Rather than simply increasing reward, it conditions your brain to prioritize the artificial surge over everything else, training the system to treat that spike as the only signal worth chasing.

That's what makes it so dangerous. It doesn't feel bad; it feels too good to stop.

The Experience: Razor-Sharp, Burning Fast

When people say cocaine makes them feel like a god, they're only half joking. The high is intensely focused on agency, movement, decisiveness, and self-assuredness. There is no psychedelia here, no mysticism. It sharpens the ego rather than dissolving it. Users feel articulate, energized, unassailable. Social interactions become rapid-fire, witty, and bold. It's easy to mistake the confidence for clarity, and in the moment, the distinction doesn't feel important.

The onset is quick, within seconds if smoked or injected, within minutes if snorted. A brief tingling or numbing of the mouth and throat is often the first signal. Then comes the rush: a surge of pleasure, a quickening of thought, a feeling that the room just got smaller and you just got larger. Time compresses. Speech accelerates. Fatigue dissolves into restless alertness. Everything feels not just manageable but obvious.

That peak is short-lived, often gone within 15 to 45 minutes. When it goes, it doesn't taper so much as drop. What follows isn't calm or even neutral. It's a negative space: irritability, a creeping anxiety, a hollow feeling that's difficult to name and harder to sit with. The conversation that felt electric a half hour ago now feels effortful. The room feels louder and smaller in a different way.

This is usually where redosing starts, not to get back to the high exactly, but to get away from the low. That distinction matters. Many people aren't chasing the rush so much as they're outrunning the crash.

The Hidden Dangers of Adulteration

Street cocaine is rarely pure. In fact, most of what's sold contains only 20 to 30% actual cocaine. The rest is made up of impurities from manufacture and *cutting agents*. These cuts are substances added to stretch the product, mimic its effects, or increase profits. Some of these are relatively harmless. Others are anything but.

Common cuts include:

- **Inert substances** like **sugar, talc, baking soda, corn starch,** or **vitamin C**, which increase weight but can be dangerous if injected
- **Caffeine** or **amphetamine**, which add stimulation and mask low purity
- Local anesthetics such as **lidocaine**, **benzocaine**, or **procaine**, which mimic cocaine's numbing effect
- **Aspirin**, which thins the blood and may be added to alter texture or taste
- **Levamisole**, a veterinary dewormer that increases the rush but can suppress the immune system and cause white blood cell disorders
- **Fentanyl**, a dangerous opioid, increasingly seen as an adulterant

Some dealers even add chemicals to make the product pass common purity tests (like the "bleach test") or to give it that numbing gum effect people associate with the real thing. A coat of hairspray can enhance the rainbow "fish-scale" sheen users look for. These additives do nothing for your high, but they can increase the risks of infections, allergic reactions, and long-term toxicity.

The worst part is, you can't tell by looking. Appearance, smell, and taste are not reliable indicators. A powder that looks clean might still contain substances that can harm or kill you.

More insidious is the increasing appearance of fentanyl in cocaine supplies. While often associated with heroin, fentanyl contamination has reached stimulant markets as well. Even trace amounts can kill. Many overdoses from "cocaine" are actually accidental opioid poisonings, unseen and immediate. Fentanyl test strips and reagent test kits, like those from DanceSafe or Bunk Police, can help identify cocaine and flag dangerous cuts or additives. The only safeguard is testing every batch, every time.

Lethal Combinations: What Not to Mix

Cocaine is inherently risky, but the real danger often lies in what it's taken *with*. In clubs, parties, and festivals, cocaine is frequently used alongside alcohol, MDMA, benzos, and other substances that interact in toxic, sometimes fatal ways. The human body can only balance so many competing signals, and cocaine is already near the redline.

Here's what not to mix.

Cocaine + Alcohol = Cocaethylene

The combination of cocaine and alcohol doesn't just compound stress on your body. It creates a new compound altogether: *cocaethylene*. Formed in the liver, cocaethylene is longer-lasting than either drug and significantly more cardiotoxic. It raises the risk of seizure, heart attack, and sudden death. Early research pointed to an 18 to 25 times greater risk of sudden death compared to cocaine alone. Not only is it more toxic, it can stay in the body up to 2 to 5 times longer, causing additional damage for a longer time. Many fatal cocaine-related incidents involve alcohol, not as a sidekick but as a catalyst.

What makes this pairing even more deceptive is how socially normalized it is. People drink to take the edge off the come-down, or they use cocaine to stay sharp while they drink, and the two behaviors reinforce each other in a loop that feels controlled right up until it isn't.

Alcohol slows your ability to recognize danger, masks early signs of cardiovascular strain, and encourages redosing far past the point where your body can compensate. By the time the warning lights would normally be obvious, the cocaethylene is already building, the heart is already working overtime, and the subjective sense of "I'm fine" no longer reflects physiological reality.

Worse, the lift you feel from cocaethylene only deepens the illusion of safety. The buzz is warm, confident, and energizing, and it convinces you that nothing is wrong even as your cardiovascular system moves into dangerous territory. The internal damage stays quiet while the high stays loud. This makes cocaine and alcohol one of the most common and most underestimated lethal combinations in social settings.

Cocaine + Fentanyl = Accidental Overdose

You don't need to *mix* cocaine with fentanyl to be at risk. These days, the risk is accidental. Street cocaine is frequently contaminated during production, packaging, or transport. A few grains of fentanyl in a line of cocaine are enough to shut down breathing, especially in users with no opioid tolerance. Deaths from "cocaine overdose" often turn out to be undetected fentanyl poisonings. This is why fentanyl test strips are non-negotiable. Every batch. Every time.

Cocaine + MDMA = Serotonin Overload, Cardiovascular Collapse

At first glance, cocaine and MDMA may seem like natural party companions. They show up in the same scenes, produce some of the same surface effects, and get combined constantly for exactly those reasons. That's the problem; mixing them is one of the most physiologically stressful combinations possible.

Both drugs increase heart rate, blood pressure, and body temperature. Together, they create a multiplicative strain on the cardiovascular system, not a simple additive effect. This can manifest as sudden heart issues, even in people with no prior history of them. Cocaine's sharp sympathetic activation combines with MDMA's empathogenic stimulation to produce a state where thermoregulation fails, dehydration accelerates, and heart arrhythmia risk spikes.

Worse, the mix can lead to serotonin syndrome. This is especially likely if the MDMA is re-dosed or misrepresented (e.g., DXM or PMA instead), or if either substance is impure.

This combo is hard on your brain, harder on your heart, and dramatically increases the risk of both short-term collapse and long-term neurochemical damage.

Cocaine + Other Stimulants = Sympathetic Overload

Combining cocaine with other stimulants (such as methamphetamine, Adderall, or even high doses of caffeine) results in compounded strain on your central nervous system. All of these drugs activate the sympathetic branch, putting the body in a state of alert, constricting blood vessels, and accelerating heart rate.

The margin for error narrows. You may feel powerful, focused, even euphoric, but beneath that, the cardiovascular load climbs. Seizures, arrhythmias, or sudden collapse are not abstract possibilities; they are well-documented outcomes of this cocktail.

Cocaine + MAOIs = Hypertensive Crisis

MAOIs include some prescription antidepressants and also appear in certain plant-based preparations like ayahuasca and Syrian rue. Many users are unaware that these interactions even exist. The result is fatal ignorance.

Monoamine oxidase inhibitors (MAOIs) interfere with the body's ability to metabolize neurotransmitters like dopamine, norepinephrine, and serotonin. When taken with cocaine (which blocks their reuptake), this creates a perfect storm. Neurotransmitter levels skyrocket, but the usual brakes are disabled.

This can lead to a hypertensive crisis: a medical emergency involving sudden, extreme spikes in blood pressure, severe headaches, stroke, or death.

Cocaine + Benzodiazepines = False Security

This is less about toxicity, more about deception. Benzos (like Xanax, Valium, or Ativan) can take the edge off a cocaine high or comedown, but they also mask warning signs. Users feel balanced, but their vitals are still under stress. The result is often unconscious overdose escalation, or dangerous behavior that would otherwise feel too risky.

Repeated pairing of cocaine and benzos can also trigger rebound anxiety, withdrawal symptoms, and in some users, dependence on both. If benzos are used to cushion the crash, use sparingly, and never with alcohol.

The Rule: Don't Stack Risk

Cocaine is already high-stakes. It speeds up systems your body depends on to stay regulated: heart, lungs, brain, mood. Stacking other substances on top, especially those that disrupt metabolism, block awareness, or add stress, pushes those systems toward failure.

Most bad outcomes aren't because someone took "too much." They happen because someone took too much of the wrong combination. If you're going to use cocaine, don't mix it with other drugs. Test everything. Space out your use. Don't stack stimulants. Avoid mixing with alcohol. Be especially wary of combinations that feel "normal" in social scenes. They're often the most dangerous ones on this list.

Harm Reduction: Using Cocaine with Care

Cocaine doesn't forgive sloppy use. Its fast onset, short half-life, and compulsive redosing pattern make it easy to overdo, and the margin between a good time and a bad outcome is narrower than most people expect. For those who choose to use it, thoughtful preparation and structured aftercare can make the difference between a rough night and something lasting, or fatal.

Even taking small steps can make a difference. You don't have to overhaul your relationship with a substance to reduce the chance of something going wrong. Small adjustments in how you prepare, who you're with, and how you pace yourself all add up.

Harm reduction is about survival. It's about recognizing risk honestly and responding with care.

Always Test Your Product

Cocaine is rarely pure. In today's unregulated market, it's not just cut with benign fillers. It's often contaminated with levamisole, phenacetin, caffeine, and increasingly, fentanyl.

Use fentanyl test strips on every batch. (See the appendix of this book for instructions on how to use a test strip.) If it's positive, do not use it. Even trace fentanyl (too small to see) can kill.

Reagent testing kits can also help detect other adulterants, although they are less definitive. If you're using regularly, they're worth keeping on hand.

Use Clean Tools and Surfaces

Don't share snorting devices. Shared straws, keys, or bills carry risk of hepatitis C, *H. pylori*, and other infections. Use individual tools, and wipe surfaces before and after. Better yet, use a small, dedicated mirror or tray that isn't passed around.

Avoid using dollar bills as straws. Small denominations are notoriously dirty, but even large bills can carry invisible bacteria. Almost every bar, restaurant, or cafe has drinking straws, often individually wrapped. If you anticipate needing more than just a few, straws are sold at every grocery store and can be trimmed to your preferred length.

Consider pre-loading lines or using a "snuff bullet" instead of dipping into a bag repeatedly. This minimizes contamination and makes it easier to monitor how much you're using.

Care for Your Nose

Space out sessions. Your nose needs days (not hours) to recover from even moderate use. Cocaine causes vasoconstriction and local anesthesia, which damages the sensitive mucosal tissue in the nose. Over time, this leads to ulcers, scabs, and septal perforations.

After using, rinse your nasal passages with sterile saline spray or a neti pot with warm distilled water. Avoid blowing your nose aggressively. Applying vitamin E oil or sesame oil (Nasya oil) to the inner nostrils can help moisturize and support healing between sessions.

Hydration, Food, and Body Awareness

Cocaine suppresses appetite and thirst, leading to dehydration and electrolyte imbalance. Drink water, but also replenish salts: electrolyte drinks like coconut water or rehydration powders are ideal. Have nutrient-dense food available for when the high fades.

Your body will often signal when you're pushing too far, but cocaine makes those signals easy to ignore. A racing pulse or tightness in the chest may feel like intensity rather than strain, and hunger can vanish long before your body has the fuel it needs. Treat these cues as early warnings rather than background noise. If you catch yourself redosing just to stay level, step back, breathe, hydrate with electrolytes, and give your system a moment to reset. Even brief pauses help your heart and nervous system recover and reduce the chances of sliding into a spiral that's harder to climb out of.

Supplementation and Neurochemical Recovery

After using cocaine, the body is often depleted of dopamine precursors and essential nutrients. To support recovery:

- **L-Tyrosine (500-1000 mg)**: Supports dopamine replenishment. It's also found in hard cheese, meat, and eggs.
- **B-complex vitamins**: Especially B-6, B-9 (folate), and B-12 for neurotransmitter synthesis and adrenal support.
- **Magnesium glycinate (200 mg)** *(not Citrate!)*: Eases tension, calms the nervous system, and supports sleep.
- **Omega-3 fatty acids**: Reduce neuroinflammation, stabilize mood.
- **Vitamin C (1000 mg) and NAC (600 mg)**: Antioxidants that help buffer oxidative stress caused by stimulant metabolism.
- **Avoid 5-HTP after cocaine**: it is more relevant to MDMA recovery and can disrupt balance when serotonin wasn't the dominant neurotransmitter impacted.
- **Sleep:** Sleep is medicine, but it may not come easily. Cocaine disrupts sleep architecture. If rest is elusive, gentle aids like melatonin, L-theanine, or magnesium glycinate can help without the rebound effects of stronger sedatives.
- **Take a break:** Give yourself some sober time between use. Avoid benzodiazepines, like Xanax. While they can blunt comedowns, they increase your risk of dependence and dangerous interactions, and can mask the signs of trouble.

Integration and Emotional Care

Cocaine gives the illusion of power, but it often leaves users physically drained, emotionally raw, and neurochemically imbalanced. The crash is real. Dopamine depletion leads to apathy and low mood. Serotonin disruption can bring anxiety, and sleep debt only compounds the damage.

The first step after use is to stop the spiral. Let it end. Don't chase a dying high into a deeper low.

Equally important is the psychological dimension. Cocaine often inflates self-perception during the high and shatters it after. This emotional whiplash can lead to guilt, self-loathing, or compulsive redosing. Recognize this for what it is: not a moral failing, but a predictable outcome of a pharmacological cycle. The emotional aftershock can be disorienting: what felt like momentum the night before may feel like emptiness the next day. This is neurochemical, not personal. Let it pass. Make a plan.

The days after use can be a useful window, if you let them be. The contrast between the high and the comedown is information. If you notice the same patterns repeating, the same triggers, the same rationalizations, the same morning-after inventory of what you said or spent or promised, that's worth sitting with. Not as self-punishment, but as data. Integration means looking at the experience honestly and deciding what, if anything, you want to do differently next time.

Talk to someone. Take a walk. Eat real food. Journal if you're inclined. Reflect on what you got out of the experience. Be kind to yourself, and reach out to someone you trust if the weight of it is more than you want to carry alone.

A Final Note

Cocaine doesn't leave bruises. What it leaves is more subtle, worn into your nervous system a little at a time. It sharpens you up, then leaves you hollow, and it's asking for more before the first rush has faded. If you're going to use it, treat it like the serious drug it is.

There's a lot to be said for the boring parts: weighing your dose, keeping your supply clean, giving yourself time to recover. Every time you do that, you're refusing to hand the drug the wheel. And it wants the wheel. That pull to keep going, to stop counting how much and how often, is baked into the chemistry. Cocaine is very good at making you forget that you're still the one deciding.

Use less than you want to. Put real space between sessions. Stay honest with yourself about where things stand. If it's costing you more than it's handing back, noticing that isn't weakness. That's you thinking clearly, which is the one thing the drug keeps trying to take.

LSD – "Acid"

"If I had a world of my own, everything would be nonsense."

The History of LSD: A Long Strange Trip

LSD-25 wasn't meant to change the world. It began, like many scientific revolutions, as a footnote in a pharmaceutical company's research log.

In 1938, Albert Hofmann, a chemist working for Sandoz Laboratories in Basel, Switzerland, was synthesizing derivatives of ergot, a toxic fungus that grows on rye. Ergot had a long and complicated history in medicine. In small doses, it could stimulate labor or constrict blood vessels. In large doses, it caused seizures, hallucinations, and death. Hofmann's goal was to develop new circulatory and respiratory stimulants. The 25th compound in that series, *lysergic acid diethylamide*, was shelved after animal tests showed little promise.

Five years later, Hofmann had what he later called a "peculiar presentiment" and decided to resynthesize LSD-25. This time, the molecule didn't stay in the test tube. After accidentally absorbing a small amount through his fingertips, Hofmann experienced the world's first acid trip: a sudden shift in consciousness marked by strange visual distortions, restlessness, and a sense of dreamlike displacement.

Three days later, on April 19, 1943, he intentionally ingested 250 micrograms, a dose now considered high (about 2.5 standard doses). He bicycled home through what would become a legendary chemical storm of kaleidoscopic visuals, looping thoughts, and existential unease. That day is now celebrated as Bicycle Day, the unofficial birth of modern psychedelia.

After Hofmann's discovery, Sandoz began marketing LSD under the name *Delysid*, supplying it to psychiatrists and researchers worldwide. From the late 1940s through the early 1960s, LSD was studied as a possible treatment for everything from alcoholism to depression to end-of-life anxiety. Therapists believed it could "unblock" repressed material, catalyze emotional breakthroughs, and even simulate psychosis in healthy subjects for research purposes.

Meanwhile, in the shadows of official science, the U.S. government, through the CIA's infamous MK-Ultra program, secretly dosed unwitting soldiers, prisoners, and civilians in a series of ethically nightmarish experiments aimed at developing mind-control techniques. These efforts produced no useful intelligence but plenty of trauma and a deep cultural distrust of psychedelics that still lingers.

By the mid-1960s, LSD had left the lab and entered the bloodstream of the counterculture. Figures like Timothy Leary, Richard Alpert (Ram Dass), and Ken Kesey championed the drug not just as a therapeutic tool but as a vehicle for personal and social transformation. The Beatles dropped acid and wrote *Revolver*. Students took it in dorm rooms and open fields. Veterans took it in silent desperation. Gurus and con artists alike claimed it could unlock the divine.

Then came the crackdown.

In 1968, under the Johnson administration, the U.S. government passed the Staggers-Dodd Bill, making LSD possession illegal at the federal level (several states, including California, had already banned it beginning in 1966).

In 1970, President Nixon signed the Controlled Substances Act, which formalized LSD's classification as a Schedule I substance, defined as having "no currently accepted medical use and a high potential for abuse." The research dried up. The culture turned. And for decades, LSD became a whisper: demonized in public, revered in secret, carried in blotters tucked inside poetry books and record sleeves.

In recent years, a new generation of scientists, clinicians, and seekers has reignited interest in LSD's therapeutic potential. Controlled trials have shown its promise in treating cluster headaches, depression, anxiety, addiction, and existential distress in terminal illness. Once again, the world is learning what Hofmann discovered by accident: that a tiny molecule measured in micrograms can shift the architecture of perception itself.

Mechanism of Action

LSD hijacks the serotonin system, amplifies neural signals, and temporarily dismantles the brain's usual barriers and pathways. Calling it a hallucinogen undersells what's happening. To understand what LSD actually does, you have to look at what it's interfering with: the structure that lets you perceive the world in the first place.

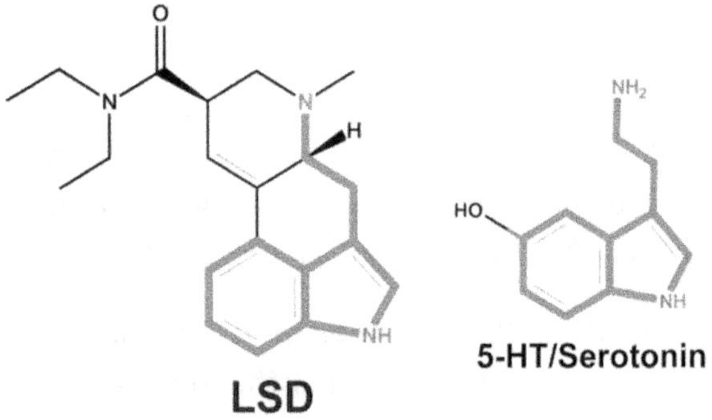

LSD and Serotonin share much of the same structure (in bold).

At the molecular level, LSD mimics and modulates the action of serotonin, the neurotransmitter involved in mood, sensory perception, and cognition. It binds with high affinity to 5-HT2A serotonin receptors, which are densely concentrated in the prefrontal cortex and the thalamus, regions responsible for abstract thinking, emotional regulation, and filtering sensory input. Unlike serotonin, LSD clings to these receptors, twisting slightly in a way that locks it in place. This conformational shift may help explain LSD's extraordinary potency and duration. A single microgram-level dose can produce effects lasting 8 to 14 hours, simply because the molecule refuses to let go.

The story doesn't stop with receptor binding. Once LSD is in the system, it begins to disrupt the brain's signaling hierarchy.

Under normal conditions, the thalamus functions as a sensory gatekeeper. It takes in a flood of raw data (light, sound, temperature, proprioception) and decides what gets passed on to conscious awareness. Most of what your senses detect never reaches the surface of your mind. The thalamus edits reality before you even know it exists.

LSD breaks that filter open.

With the gating function weakened, sensory input floods through with far less regulation. Details that are normally suppressed (shadows shifting, textures vibrating, the subtle crackle of electricity in a room) become vivid, sometimes overwhelming. The word "hallucination" doesn't capture it. What's actually happening is a reorganization of what gets attention and how it's interpreted. Background becomes foreground. The implicit becomes undeniable.

Once the feedback loop is disrupted, signals start to distort. The processing loop quickens, and gaps begin to form in your model of the world. Your brain fills them in with what it has: your thoughts, your memories, your feelings.

Inside the cortex, things get stranger. LSD increases global brain connectivity, meaning regions that don't typically communicate begin lighting up in conversation. The visual cortex starts trading data with emotional memory centers. Sounds acquire color. Thoughts loop back into sensation. This is part of why LSD often produces synesthesia, where the boundaries between senses blur and overlap.

At the center of all this is the Default Mode Network (DMN), a group of interconnected brain regions associated with introspection, self-reference, and the ongoing narrative you tell yourself about who you are. It's the internal voice that says

"I am me" and stitches your past to your present. Under the influence of LSD, activity in the DMN drops sharply. The sense of a continuous, stable identity begins to erode. This is the neurological basis for what many describe as ego dissolution: that expansive, and sometimes terrifying, sense of becoming one with everything, or nothing at all.

In this state, experience becomes unmoored from the usual boundaries of self and time. Memories, sensations, emotions, and imagination blend into a single ongoing process, often accompanied by a sense of cosmic significance. Whether you interpret that as spiritual awakening, psychospiritual disintegration, or just your brain on overdrive depends largely on context. The drug doesn't pick a direction. It opens a door.

What LSD does is alter the filtering process, changing how information is prioritized, sorted, and integrated. It turns up the gain, collapses the partitions, and forces networks into communion that are usually kept apart. There's no sedation, no numbing, no foreign content inserted into your consciousness. Everything you experience was already in the system. LSD just rearranged it.

That means LSD alters how you construct reality, not just how you feel about it. The implications can be beautiful. They can also be deeply destabilizing. Which is why, as always, set and setting matter more than the molecule.

Preparing for Liftoff

Preparation matters because LSD is not a casual substance, and the way you approach it should reflect that. You are about to peel back many of your emotional filters and walls. Treat the experience with intention and gather what you need before you start.

First things first: Test your substance.

Not all blotter is LSD. There are plenty of lookalikes (25i-NBOMe, DOC, and other research chemicals) that can feel similar at first but have wildly different safety profiles. NBOMe compounds, for example, are active at microgram levels like LSD, but unlike LSD, they can kill you. Use an Ehrlich reagent test. It takes sixty seconds and can save your life. Get one at DanceSafe or a similar site. If you're not sure, don't take it.

Have a sober sitter.

LSD can be intense, even for those experienced with it. Have someone whom you trust who can keep an eye on things. Make sure they know how much you've taken, and when, and can remain sober through the whole experience. Don't wait until you're too far gone to try to call someone. As anyone who has ever tried to use a phone while on psychedelics will tell you, this is better set up beforehand.

Don't forget, Set matters. A lot.

That means your mindset. Don't trip when you're in a dark place, going through a breakup, or trying to escape bad feelings. LSD doesn't make things better; it makes them louder.

Whatever is unresolved will be waiting for you in kaleidoscopic Technicolor, whether you planned to examine it today or not.

Setting also matters. A lot.

Where are you going to be? With whom? Is it a place you can lie down if needed? Will you be safe if you forget your name for a while? The ideal setting is somewhere familiar and emotionally safe. Some people enjoy being in nature, which is great if you have access to it and someone to make sure you don't wander off into a creek thinking you're a salmon. Indoors works too, if you'd like more control over your environment. Loud strangers, sharp objects, and unsupervised fire are all potential liabilities.

LSD distorts your sense of time so completely that you may feel like hours have passed after ten minutes, or vice versa. Knowing what time you dropped also helps you orient, if you hit a weird loop. Write it down, or start a stopwatch on your phone (and then put it away) so you can have an objective way to know how much time has passed.

Dosing: Less is often more.

While there have been no confirmed casualties from overdose on LSD alone, there have been *many* people who took too much and had a bad time. A standard dose is about 100 micrograms, but that varies depending on the blotter and your body. Start with no more than one dose unless you know the source and your personal response. LSD's effects can take an hour or more to kick in, and can last for over 14 hours. "I don't think it's working, give me another" is the epitaph to many an overdose. Don't redose trying to chase something; just ride what you've got. If you want to take more next time, you can, but there's no undo button on this kind of chemistry. **You can always take more, but you can't take less.**

Consider a parachute.

Have a benzodiazepine on hand (like Xanax or Klonopin) *if prescribed and appropriate*. Not because you plan to use it, but because it's reassuring to know there's a way out if things go sideways. Benzos can abort a trip, smooth anxiety, and bring you back to baseline. Avoid sleeping pills like Lunesta or Ambien. They're too weak and won't reliably end the trip or put you to sleep, often compounding the effects and causing a trippy sleep-zombie state that is best avoided.

LSD and Other Drugs

By the numbers, LSD is one of the physically safest drugs in this book. There is no reliable lethal dose in humans, and no confirmed case of anyone dying from the drug itself. It does not wear down an organ over the years the way alcohol goes after your liver. The danger with LSD lives almost entirely in your head, not your body. A trip can be confusing, even frightening, and that is worth taking seriously.

That easygoing reputation mostly holds when you start mixing LSD with other things. Acid is not especially reactive. Most of the time, combining LSD with another drug changes the character of the trip rather than putting your life at risk. There are a few exceptions that matter, and they are worth knowing.

The one that earns a real warning is lithium. If you take lithium, usually prescribed for bipolar disorder, do not combine it with LSD. A 2021 analysis out of Johns Hopkins gathered dozens of firsthand reports from people who had mixed the two, and close to half of them involved seizures, with many serious enough to need medical attention. This is the rare LSD interaction that can genuinely land you in a hospital, so it is not a corner to cut. Tramadol, a painkiller that lowers your seizure threshold, should also not be mixed with LSD, for similar reasons. If you are taking lithium or tramadol, skip the acid.

Stimulants are less about physical danger and more about wrecking your afternoon. Uppers like cocaine or amphetamines (or even a lot of caffeine) push you toward anxiety, and anxiety is the raw material of a bad trip. LSD already turns up whatever you walk in carrying, so a racing heart and a jittery nervous system just tilt the odds toward the version of the day nobody wants.

Some people fold a measured dose of MDMA into a trip to blend the empathy of MDMA with the insight of LSD, a pairing called *"candyflipping."* This asks for more care, not less, because the stimulant half is doing the opposite of settling you down. If you choose to mix these, start with smaller doses of both, and make sure you note the timing, as these have different durations. Wait until the LSD is already underway before adding another substance.

Antipsychotics are a stranger situation. Drugs like risperidone or quetiapine block the same 5-HT2A receptors LSD works through, so they tend to flatten or switch off the trip rather than intensify it. That is actually how emergency rooms used to bring people down from bad ones. The trouble is twofold. A muffled trip tempts people to redose, chasing an effect the medication is busy blocking, which is how you end up having swallowed far more acid than you intended. If you're trying to use these to end a trip, they often cause an emotional crash that is best left to the professionals. More importantly, if you are on an antipsychotic, you are usually being treated for something like schizophrenia or bipolar disorder, and those are exactly the conditions a strong psychedelic can destabilize or pull to the surface. The real interaction there is not chemical. It is between the drug and the mind it is landing on.

Everyday antidepressants tend to be less dramatic. SSRIs, the common ones, usually dull LSD when you're on them for a while, so plenty of people on them find acid underwhelming rather than risky. The older MAOIs are less predictable and can send the experience somewhere strange, so give those more room. None of this is a reason to quit a medication you actually need in order to chase a trip. That trade is almost never worth making.

Downers land in the middle. Alcohol or benzodiazepines will sand the edges off a trip, which sounds pleasant and sometimes is, but they also fog the judgment you most want intact while your perception is already unreliable. A few drinks into an already disorienting headspace is exactly how people wander into decisions they would never otherwise make. If you use them at all, keep it light.

The short version is simple. On its own, LSD is remarkably forgiving of your body and a good deal less forgiving of your mind. Most combinations only change the flavor of the trip, but a few, lithium and tramadol above all, carry a physical risk that no good night is worth. And whatever you put with it, remember that acid turns up the volume on whatever is already playing. The calmer and clearer you go in, the better your chances of coming out the same way.

Nutritional Neurotransmitter Support

While LSD is not known to be neurotoxic or dangerous in the way that some other substances are, the intense neurological activity it induces can lead to temporary depletion or imbalance in neurotransmitter systems, particularly those involving serotonin, dopamine, and acetylcholine. Supporting the brain's recovery through nutrition and supplementation can help reduce the post-experience "hangover" some users report, and promote overall mental clarity and stability.

One key nutrient of interest is choline. Choline is an essential nutrient involved in the production of acetylcholine, a neuro-transmitter crucial for memory, learning, and neuromuscular function. Although LSD's primary action is on the serotonin system, the acetylcholine system plays a complementary role in modulating attention and sensory processing, and is also affected indirectly during altered states of consciousness.

After an LSD experience, the brain's signaling systems often need time and resources to re-equilibrate. Supplementing with choline can support this process, particularly if one experiences brain fog, reduced focus, or mental fatigue in the days following use. Choline may help rebuild depleted reserves and restore optimal synaptic function.

Common dietary sources of choline include eggs (especially the yolk), liver, soybeans, and some fish. Supplemental forms include alpha-GPC, CDP-choline (citicoline), and choline bitartrate. Among these, alpha-GPC and CDP-choline are more bioavailable and cross the blood-brain barrier efficiently, making them preferable for cognitive support. Start with **300-600 mg**, and adjust according to your needs.

While no supplement can entirely eliminate the psychological complexity of a psychedelic experience (or substitute for proper integration), nutritional strategies can help ground and replenish the body. In addition to choline, users often benefit from:

- **Magnesium glycinate (200 mg)**, to support muscle relaxation and sleep
- **Vitamin C (1000 mg)**, for antioxidant protection
- **Omega-3 fatty acids (1-2 g EPA/DHA)**, to support neuronal membrane health
- **B-complex vitamins**, for energy metabolism and nervous system function

Final Thoughts

Strip away the myths and the mechanism, and this whole chapter comes down to one thing about LSD: it takes the filters off. For a few hours the world comes in raw and unsorted. That can be beautiful. It can also be overwhelming, which is the whole reason preparation and integration matter.

None of this is about turning your trip into a lab experiment. Test your tab, and settle where you'll be and who you'll be with, before you dose. Do your preparation and get out of its way.

That's the difference between a journey and an escape. An escape is running from where you are. A journey is choosing to go somewhere, and paying attention on the way. You're not just taking a drug. You're deciding, on purpose, to spend a few hours as a slightly different version of yourself. Pack for it like you mean it.

Psilocybin Mushrooms

"One side will make you grow taller, and the other side will make you grow shorter."

From Prehistory to Modern Day

Psilocybin mushrooms have been eaten, worshipped, outlawed, rediscovered, and decriminalized (often in that order) by nearly every culture that's come into contact with them. They have survived prohibition, colonization, and cultural amnesia.

Psilocybin is the primary psychoactive compound in dozens of mushroom species, including *Psilocybe cubensis, semilanceata, azurescens,* and others. Once you eat them, your body converts psilocybin into *psilocin*, which is the molecule that actually does the work in your brain.

Psilocin looks a lot like serotonin, your brain's mood and perception regulator, and it fits into many of the same receptors. Specifically, it activates the 5-HT2A serotonin receptor, the same one targeted by LSD and DMT. When psilocin shows up at that receptor, it slips right in.

Mushrooms aren't mini-LSD. The trip is distinct, often more emotional, more body-based, and sometimes less controllable. If LSD is a microscope, psilocybin is a mirror. Sometimes that mirror doesn't care if you're ready to look.

Psilocybin mushrooms have been used ceremonially for centuries (if not millennia), particularly in Central and South America. The Aztecs called them *Teonanácatl*, meaning "divine flesh," and consumed them for ritual, healing, and celebration. Mushroom use was common in Mayan, Mixtec, and Mazatec cultures as well. It wasn't just spiritual; it was political. Mushrooms were eaten during coronations, to commune with the gods, or to determine the future.

That changed with the arrival of the Spanish, who declared such practices heretical and associated mushroom use with demonic possession. They outlawed psilocybin ceremonies and punished indigenous use with missionary violence, but traditions endured underground, often passed orally and out of reach of European understanding.

Western science remained mostly clueless until the 1950s, when R. Gordon Wasson, a banker with a taste for ethno-botany, traveled to Oaxaca and participated in a mushroom ceremony with Mazatec curandera María Sabina. His 1957 photo essay in Life magazine introduced psilocybin to a curious, post-war American public, and to Albert Hofmann, who soon isolated and synthesized psilocybin and psilocin for Sandoz Pharmaceuticals.

Wasson's work helped catalyze a wave of interest in psychedelic mushrooms among researchers, spiritual seekers, and countercultural figures. In the 1960s, Harvard psychologists Timothy Leary and Richard Alpert (later known as Ram Dass) began the Harvard Psilocybin Project, administering the compound to students, prisoners, and religious scholars. Their Concord Prison Experiment claimed reductions in recidivism, and the Marsh Chapel Experiment famously led to profound mystical experiences among divinity students.

As the psychedelic movement spread, mushrooms were embraced by artists, musicians, and activists alike. They became associated with anti-establishment thinking, expanded consciousness, and the era's broader experimentation with altered states, but public enthusiasm triggered government backlash. Psilocybin was federally banned in the United States in 1968 and later added to the United Nations' Convention on Psychotropic Substances in 1971.

Despite prohibition, underground use continued. Terence McKenna, an ethnobotanist and writer, popularized mushroom cultivation and developed philosophical frameworks around their use, most notably his theory that psychedelic mushrooms had influenced the evolution of human consciousness. His "Stoned Ape Hypothesis," while speculative, captivated many and helped anchor mushrooms in psychedelic lore.

Throughout the 1980s and 1990s, mushroom use persisted quietly among psychonauts, festival-goers, and spiritual practitioners. Then, in the 2000s, clinical interest began to resurface. Early studies at institutions like Johns Hopkins and NYU demonstrated psilocybin's potential to alleviate anxiety, depression, and addiction. These findings helped reframe mushrooms not as recreational drugs, but as powerful tools for healing and self-inquiry.

Today, psilocybin is being decriminalized in major cities and studied in clinical trials worldwide. Oregon and Colorado have legalized supervised therapeutic use. What was once sacred, then suppressed, then underground, is becoming mainstream again. The cycle turns, one more time.

What Mushrooms Do (and How)

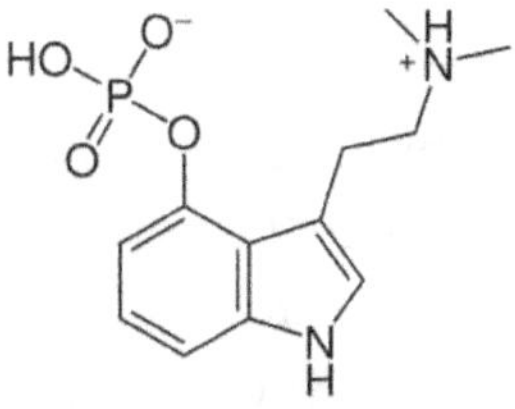

Psilocybin

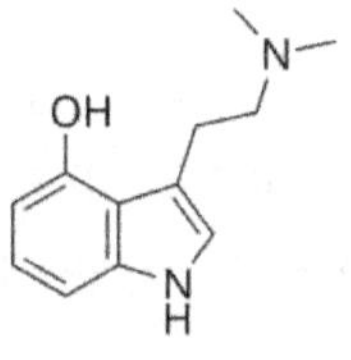

Psilocin

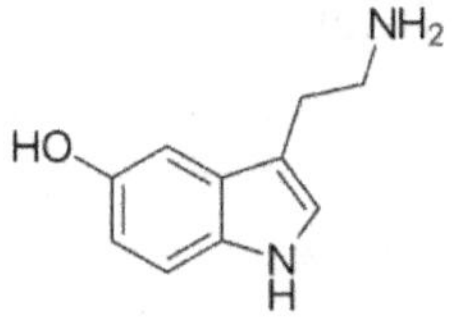

Serotonin

The active ingredients in psychedelic mushrooms are *psilocybin* and *psilocin*. When you consume them, your body converts psilocybin into psilocin, which is the molecule that actually affects your brain. Psilocin looks very similar to *serotonin*, one of the brain's most important neurotransmitters. That similarity is not a coincidence. It's the reason mushrooms can shift perception so dramatically.

Serotonin helps regulate mood, emotion, sensory input, and the way your brain responds to the world. It acts like a filter, helping your mind decide what to focus on and what to ignore. Psilocin, because of its structural similarity, can bind to many of the same receptors, especially *5-HT2A*. This is the same receptor that LSD and DMT also target. When psilocin activates it, the usual filters start to loosen.

Under normal conditions, your brain spends a lot of energy organizing and simplifying information. The world is too complex to process every detail, so your brain builds models, incorporating patterns, habits, and expectations, that help you make sense of things quickly. Psilocin disrupts this filtering system. Instead of just showing you what's useful, the brain starts showing you what's actually there. You begin to notice details that are normally invisible. Colors become richer. Patterns emerge in wood grain or clouds. Emotions surface that you may not have fully felt in years.

Internally, psilocin causes regions of the brain that don't usually communicate to start talking to each other. At the same time, it reduces activity in something called the *default mode network,* which is a group of brain areas involved in self-reflection, memory, and identity. This network is what helps maintain your sense of being a separate "you" moving through the world. When its activity is reduced, the boundaries between self and environment can begin to dissolve. This is where the feeling of ego loss or unity with everything often comes from.

Psilocybin reorganizes the brain rather than shutting it down. Imaging studies show that during a mushroom trip, the brain becomes more globally connected. You may experience *synesthesia*, where sounds have color, or thoughts seem to echo as physical sensations. This is not hallucination in the cartoon sense. You are not seeing things that aren't there. You are experiencing familiar things in unfamiliar ways.

Another key part of mushrooms' effects comes from the way psilocin affects *glutamate*, the brain's main excitatory neurotransmitter. Glutamate is responsible for learning, memory, and mental flexibility. When psilocin interacts with serotonin receptors, it indirectly influences glutamate activity. This can produce feelings of insight, emotional breakthrough, or renewed perspective.

Compared to other psychedelics like LSD, psilocybin is shorter-acting and usually less sharp or analytical. It tends to feel more organic, more emotional, and more bodily. While LSD might send your thoughts racing through abstract geometry, mushrooms are more likely to drop you into your heart or your gut. The insights can be profound, but they often come in the form of emotion rather than logic.

Every brain responds differently, but the mechanism remains consistent. Psilocin binds to serotonin receptors, disrupts the usual patterns, and opens the door to a different way of seeing, thinking, and feeling. That door does not always lead where you expect. Sometimes it takes you inward. Sometimes it brings buried material to the surface. And sometimes it just shows you how strange and beautiful the world already is.

The mechanism is consistent, but the experience is personal. These substances open a window. What you see through it depends on what you've been carrying, and what you're willing to set down. The mushroom doesn't care either way. It just keeps the window open until the chemistry wears off.

How to Not Feel Like You're Dying

Psilocybin mushrooms have a strong safety record. Nobody has died from them, and the physiological risk is close to zero. Your digestive tract doesn't care about that statistic. The most common physical complaints are nausea, gas, stomach cramps, and diarrhea, and they catch more beginners off guard than you'd expect. Your body is reacting to *chitin*, the tough, non-digestible fiber that makes up the mushroom's cell walls. Human enzymes barely touch it, so it sits in your stomach being difficult while everything else gets absorbed around it. Evolution didn't design the human gut with moldy forest floor fungi in mind, even if they do open your third eye. None of this is dangerous. Most of it has a workaround.

Dosing and Timing

Potency varies between species, between batches from the same grower, and even between the cap and stem of a single mushroom. Stick to a known, dried source, ideally *Psilocybe cubensis*, and weigh your dose on a scale accurate to at least a tenth of a gram. Eyeballing dried mushrooms is a reliable way to end up somewhere you didn't intend to go.

For most people, 1.5 to 2.5 grams of dried *Psilocybe cubensis* is enough for a full, meaningful trip. That range isn't a floor to build from. It's where the work actually happens. A threshold dose, light enough to notice the shift without losing your footing, starts around 0.25 grams. Nearly all the intentional, productive mushroom experiences people describe happen somewhere between that threshold and 2.5 grams.

Doses above 3.5 grams exist, and so do doses of 5 grams or more, sometimes called "heroic doses." Neither is a goal. Both belong to a different category of experience, one that demands more preparation, more support, and a more honest look at whether you're ready for that level of dissolution. Most people aren't, and most people don't need to be. The insight you're after is unlikely to be waiting for you at a higher number on the scale.

Whatever you take, give it time. Onset typically runs 30 to 60 minutes. It takes longer if you've eaten recently, since food in your stomach slows absorption; an empty stomach speeds things up and sharpens the come-up. Chewing or grinding your mushrooms thoroughly helps too, since it starts breaking down the chitin mechanically before your stomach has to do it alone. Wait at least two full hours before considering a redose, and understand that by the time redosing crosses your mind, the first dose is usually still arriving. Patience isn't optional here. **You can always take more. You can't take less.**

Body load is the physical side of a mushroom trip, and it often shows up before any of the perceptual effects do. Waves of nausea. A heavy feeling in your limbs, like gravity got turned up a notch. Your stomach may cramp or churn, your skin might flush or run cold, and your heart rate typically climbs. None of this means anything has gone wrong. It's your body processing an unfamiliar compound and a stomach full of solid organic material it doesn't know what to do with. For most people, body load peaks somewhere in the first 30 to 60 minutes and eases off as the mushroom clears your gut and the visual and emotional effects take over. It can be intense enough to make people think something is physically wrong, and that misread is where a good share of "bad trip" panic actually starts, not from anything you're seeing, but from unfamiliar bodily sensations that get mistaken for something more serious. Recognizing body load for what it is, a temporary and uncomfortable digestive process rather than a medical emergency, takes a lot of the fear out of it. There are also concrete ways to blunt it before it starts.

Mushroom Tea

One of the most effective ways to reduce body load is to not eat the mushroom solids at all. Grind the mushrooms and steep them in hot water with a splash of lemon juice and some ginger or peppermint. A peppermint tea bag works fine. Let it sit for 10 to 20 minutes until the water turns golden yellow, then strain out the solids. You can pour a second round of hot water over the spent grounds and get a second batch. Toss the solids after that.

The lemon juice helps convert psilocybin to psilocin before it even hits your stomach, which can speed up onset slightly. The ginger and peppermint are great at both covering the taste and settling your stomach. Most people find the nausea either disappears or becomes much more manageable this way.

Lemon Tek

The procedure known as "*Lemon tek*" goes a step further. Grind your mushrooms as fine as you can and submerge them in fresh lemon juice (just enough to cover them) for 15 to 20 minutes, stirring occasionally. The citric acid does the same conversion work as above, but more aggressively. Strain the solids, drink the liquid, and chase it with water if the taste is too much.

The tradeoff is that lemon tek tends to hit harder and faster, and it compresses the trip into a shorter window. That's not always better. A faster, more intense come-up is harder to navigate, especially if you haven't spent much time with mushrooms. If you're newer to this, the tea is a gentler option. Lemon tek is a tool, not an upgrade.

Don't Stack Without Thought

While mushrooms don't have many dangerous interactions, some combinations can make things harder:

- **Alcohol**: Can intensify nausea, blunt insight, and add to dehydration. Not dangerous, but usually doesn't add anything good to the experience.
- **Tobacco**: Can hit harder than expected. Compounds found in tobacco smoke inhibit MAO activity and may intensify the trip unpredictably. If you're having a bad time, that cigarette may not help. Note that this only applies to smoked tobacco, not nicotine patches, vape juice, pouches (like Zyn), or gum.
- **Cannabis**: Wild card. Can soften or sharpen the experience. If in doubt, wait until the peak has passed.
- **LSD & 2C-B**: Cross-tolerance can reduce the effects of one or both. Space them out by at least 48 hours at a bare minimum, with 10-14 days for a full reset.
- **MAOIs/Ayahuasca:** MAOIs, such as those prescribed for depression, or those used with Ayahuasca, can produce dangerous and unpredictable effects, including serotonin syndrome or hypertensive emergency.
- **Lithium/other Antipsychotics:** Lithium and other anti-psychotics are dangerous to mix with mushrooms, both because of the documented risk of seizures, and because of the tendency of mushrooms to aggravate the underlying conditions antipsychotics are designed to treat.
- **Stimulants:** Cocaine, amphetamines, and even caffeine, while not dangerous, can amplify the trip in unpredictable ways. Some people choose to combine mushrooms with MDMA (sometimes called *hippy-flipping*). While this combination is not dangerous, take less of each of them until you get a feel for how they affect you.

Supplements and Recovery

Psilocybin doesn't deplete serotonin like MDMA, but it can leave you feeling mentally raw. Some people find that these help:

- **Choline (Alpha-GPC or CDP-Choline) (300-600 mg)**: Restores cognitive clarity and reduces mental fatigue.
- **Magnesium Glycinate (200 mg)**: Eases physical tension and improves sleep.
- **Omega-3s (1-2 g EPA/DHA)**: Reduce neuroinflammation and support mood.
- **Vitamin C (1000 mg)**: Antioxidant support, especially if sleep was poor.

The effects may end after a few hours, but their impact can linger for days or weeks. Give yourself time to process. Journaling, walking, art, and conversation with trusted people can help you make sense of what happened.

Sit with it. If it was pleasant, what made it that way? If it was unpleasant, what can you learn from it? Many times, more can be learned from a hard trip than a giggly one.

Tolerance to psilocybin builds quickly and drops slowly. Taking mushrooms again within a few days will feel muted. A good rule: wait at least two weeks, preferably longer. Give your brain, and your story, time to settle.

Not All Mushrooms Are the Same

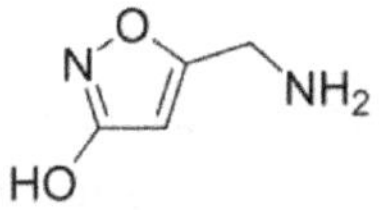

≠

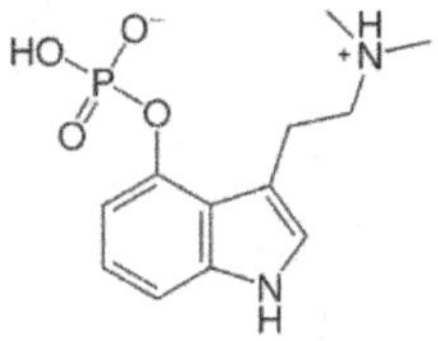

Muscimol *Psilocybin*

Not all psychoactive mushrooms contain psilocybin. Fly Agaric (*Amanita muscaria*), the red-and-white mushroom found in fairy tale illustrations, contains *muscimol*, not psilocybin. Its effects are more sedative and deliriant, often dreamy, intoxicating, and confusing. While Fly Agaric has its own rich cultural legacy, it's not the same mushroom or trip and should not be treated as interchangeable.

Some smoke shops and retailers have begun selling Amanita-based products under the label of "legal mushrooms." Be careful! They can produce strong, unpredictable effects, and they are not substitutes for psilocybin. Check the label when considering purchasing them so you know what you're getting.

Note also that these products are unregulated, and have been linked to salmonella outbreaks and heavy metal poisonings due to tainted products. Some have been found to contain research chemicals or psychedelic analogs. No one is testing them. There's truly no guarantee that the product inside contains what it claims, no matter how slick the label looks.

Reverence Is Not a Plan

Psilocybin has been called sacred for about as long as people have been taking it, and for many that reverence is real and earned. It's no substitute for preparation. A reverent state of mind won't fix a misjudged dose, won't rehydrate you, and won't do much for a setting that turns hostile two hours in. Your body still runs on serotonin, on what's in your stomach, and on how much you slept. None of that bends to how profound the trip feels.

Two things decide most of it: how well you prepared, and how honestly you're willing to look at what comes up. A well-prepared dose can surface what you've been avoiding, which is usually the point, though it tends to do that on its own schedule rather than yours. Better to make that bargain on purpose than stumble into it.

Prepare like it counts. Weigh the dose instead of eyeballing it, settle your head before you start, keep a clock and a sober friend within reach, and leave yourself room afterward to come down. The hours on mushrooms are the easy part. What you do with them in the unremarkable days after is harder, and it's where most of the value ends up.

Ketamine

"I can't explain myself, I'm afraid, Sir, because I'm not myself you see."

Introduction: A Drug with Two Faces

Ketamine is a substance of contradiction. It's used to sedate children in emergency rooms and to induce transcendental experiences in underground ceremonies. It's both a surgical anesthetic and a club drug. And in recent years, it has re-emerged as one of the most promising treatments for major depression.

This dual identity, clinical and chaotic, healing and hedonistic, makes ketamine one of the most compelling molecules in modern psychopharmacology.

History and Medical Use

Unlike many familiar substances, ketamine didn't come from ancient rituals, passed-down wisdom, or gifts from sacred plants. It actually started with the search for a better anesthetic. In the early 1960s, researchers were looking for a safer alternative to *phencyclidine* (better known to D.A.R.E. graduates as *PCP*), which had shown potential as an anesthetic but quickly gained a reputation for causing violent delirium and psychosis. Chemists at Parke-Davis Laboratories, led by Calvin Stevens, synthesized ketamine in 1962. They were hoping to create something less extreme. What they discovered was a molecule that could produce surgical anesthesia without shutting down breathing. It worked quickly, reliably, and safely, and it changed everything.

Initial testing began on animals, then moved to human subjects, including prisoners at Jackson Prison in Michigan. Patients described unusual psychological effects. Some felt like they were floating. Others said they left their bodies entirely. Despite the strangeness, ketamine proved incredibly useful. It was adopted rapidly in hospitals and clinics.

Its first major use came on the battlefields of Vietnam. Medics needed something that worked fast, and was effective and easy to administer. Ketamine could sedate wounded soldiers without expensive equipment or constant monitoring. It left breathing intact, which meant it could be used in chaotic or low-resource environments. Soldiers called it the "buddy drug," since it could be administered by a fellow soldier in the field. For many, it was the last thing they remembered before surgery, and sometimes the last moment of peace before waking up in a hospital far from the jungle.

After Vietnam, ketamine spread through civilian medicine. It became a staple in pediatric emergency rooms and veterinary clinics. Doctors appreciated its quick onset and wide safety margin. Vets appreciated how well it worked on cats, dogs, and large animals. That is how it picked up the labels "cat tranquilizer" and "horse anesthesia," nicknames that would later follow it into party culture.

By the late 1970s and early 1980s, ketamine began to slip into the underground, where it started doing what it was originally built to do: replace PCP. By that point, PCP had earned a brutal reputation. The highs were unpredictable, the aggression was real, and the horror stories had piled up enough to scare off even adventurous users. Ketamine offered a similar dissociative territory without the violence or the multi-day fallout. Psychonauts and experimental therapists started exploring its effects outside clinical settings. It was sometimes compared to LSD, though the experience was very different. Where LSD brought visions of unity and divine light, ketamine brought silence, distance, and detachment. For some, that made it even more powerful.

In the 1990s, ketamine entered the club scene. It showed up at raves in the United Kingdom, in underground parties in Hong Kong, and eventually in the U.S. In powder form, it was often sold as or mistaken for MDMA. This led to unexpected outcomes. Someone expecting to feel euphoric and connected might instead collapse into silence and lose contact with their body. Some people found this terrifying. Others found it fascinating. Ketamine became known as a short, strange trip, easy to control but hard to explain.

During the 2000s, concerns about ketamine began to rise. In China and other parts of Asia, recreational use surged, with some users going through entire boxes of vials in a week. Long-term use was linked to severe bladder damage, memory problems, and compulsive redosing. Governments began to restrict access. In 2006, ketamine became a controlled substance in the United Kingdom. Other countries followed.

At the same time, researchers were rediscovering ketamine's medical value. Small clinical trials in the early 2000s showed that low doses could rapidly relieve depression, even in patients who had failed every other treatment. The effects were stabilizing, not psychedelic. Patients reported a sudden drop in suicidal thoughts. They felt clear, calm, and able to cope, sometimes within hours. This sparked a wave of research into ketamine as a psychiatric medicine.

By the 2010s, clinics offering ketamine infusions began to appear in cities across the United States. Some were run by psychiatrists. Others looked more like spas. The FDA approved a nasal spray version called esketamine in 2019, intended for treatment-resistant depression. Insurance companies began to cover it. What was once a battlefield anesthetic had become a mental health tool.

Meanwhile, in the underground, ketamine was evolving again. Ketamine-assisted psychotherapy became more structured. Some facilitators adapted ceremonial frameworks, borrowing elements from other psychedelic traditions. Sessions included music, guided reflection, and integration work. The experience was framed as an opportunity for healing and insight, with preparation and follow-through built into the process.

Today, ketamine sits at a cultural crossroads. It is used in emergency rooms and at forest gatherings. It appears on the World Health Organization's list of essential medicines. It also shows up in festival kits, shared in balloons or bumps behind the DJ booth. It is used carefully and carelessly, in hospitals and at parties, for healing and for escape.

What makes ketamine unique is that its chemistry has never changed. Only the stories around it have. It has been used to numb pain, to dissolve identity, to stabilize minds, and to escape reality. It offers space. What you do with that space is up to you.

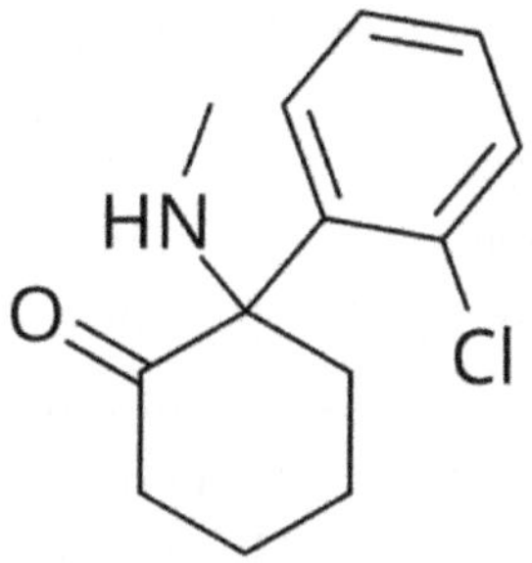

Ketamine

How It Works

Most people think of drugs as substances that either speed you up or slow you down, but ketamine doesn't fit neatly into that model. It doesn't make you sleepy in the traditional sense, and it doesn't make you feel high like stimulants or opioids. What it does is something far stranger. It severs the connection between your body, your sense of self, and your usual patterns of thought. This effect is called *dissociation*, and ketamine is one of the few substances that can induce it so reliably.

At the molecular level, ketamine works by blocking the *NMDA receptor*. NMDA receptors are part of the *glutamate system*, which is the primary excitatory network in your nervous system. If dopamine is about reward and serotonin is about mood, glutamate is about awareness. It handles how your brain processes the world in real time.

When ketamine blocks NMDA receptors, that processing starts to break down. Imagine trying to stream a video over a dial-up connection. The image stutters. The sound goes out of sync. Eventually, the screen goes black, or freezes on a single frame. That's what your brain does when it can't process its normal inputs. The result is a kind of sensory and cognitive disconnection, a suspension of ordinary reality.

This is where dissociation begins. Your thoughts feel distant. Your body may seem foreign. Time loses meaning. Some people feel like they're floating above themselves. Others experience total ego loss, where identity, memory, and even language fall away. This is what people mean when they talk about "falling into the K-hole." It's not always pleasant, but it's not random noise either. There's structure to the silence.

Ketamine also reshapes the signals it blocks.

When NMDA receptors are shut down, other systems start to compensate. One of the most important is the AMPA receptor, another part of the glutamate network. Activating AMPA appears to trigger a surge in brain-derived neurotrophic factor, or BDNF, a chemical that supports neuron growth, repair, and plasticity. This is believed to be one of the reasons ketamine has such fast-acting antidepressant effects. It may help your brain rewire how it processes pain, which is a different mechanism entirely from dulling it.

Ketamine also nudges the dopamine system, which explains the temporary feelings of pleasure or motivation. It also touches the opioid system, which may be why low doses feel calming and safe. Still, its primary identity is dissociative, acting as a tool for disconnecting you from your usual sensory, emotional, and psychological patterns.

In low doses, this can create a gentle sense of detachment or floating. In moderate doses, it can lead to deep introspection or emotional reset. In high doses, it can dissolve your connection to your body, your environment, and even your identity. That sounds scary to some, and sacred to others. The setting makes the difference.

What makes ketamine so unusual is that it can work as a surgical anesthetic, a treatment for depression, and a party drug, all depending on the dose, intention, and context. The same mechanism that numbs pain can also reveal hidden trauma. The same disconnection that makes it popular in clubs can also allow for healing in therapy.

Ketamine removes the scaffolding your mind normally relies on and lets you experience what's underneath. The signal comes back. It always does. The question is whether you learned anything while it was gone.

Subjective Effects and the K-Hole

Ketamine's effects depend heavily on dosage. At low doses, users may feel floaty, dreamy, or mildly euphoric. Visual distortions emerge: tracers, doubling, shifting perspectives. Mid-level doses create a deeper dissociation where the body seems far away and time loses meaning.

At high doses, ketamine can induce what's colloquially known as the "K-hole." In this state, users may lose all sense of self and body, experience visions, or perceive alternate dimensions. For some, it is mystical or expansive. For others, it is dark and terrifying. Context, mindset, and intention matter greatly.

Therapeutic Applications

In clinical settings, ketamine is administered in low, sub-anesthetic doses to treat depression. Unlike SSRIs, which may take weeks to work, ketamine can relieve symptoms, including suicidal ideation, within hours. This has made it a lifeline for those with treatment-resistant depression.

Ketamine is also being studied and used off-label for PTSD, anxiety, and chronic pain. Ketamine-assisted psychotherapy combines the drug with guided therapeutic sessions. Here, the dissociation creates space, allowing patients to revisit trauma without becoming overwhelmed.

If you're considering ketamine for therapeutic use, a quick internet search can help you find a legal clinic, either in your area, or online.

Note that the therapeutic value doesn't come from the drug alone. *Integration*, what you do with it afterward, is where the change takes root.

Harm Reduction

Ketamine isn't your average party drug, and it isn't your average antidepressant. It's surgical in the right hands and unpredictable without them. If you're going to use it, recreationally or therapeutically, you need to treat it with respect and a plan.

Bladder Care

Bladder damage is one of the real risks that separates ketamine from most other psychedelics. It doesn't happen to everyone, but when it does, it hits hard and it lingers. Long-term, high-frequency use can lead to ketamine-induced cystitis, marked by chronic inflammation, pain, and sometimes permanent damage. You'll know it by the burning, urgency, and that horrible feeling of always needing to pee but never being done. It's not rare, and it's not reversible.

If you're going to explore with K, hydrate smart and urinate regularly. Don't just lie in a puddle of euphoria and forget your organs. Space out your sessions. Your bladder is not a disposable item.

N-Acetylcysteine (also commonly called *NAC*) is one of the most useful supplements in harm reduction, and one of the least talked about. It supports glutathione production (your brain's internal cleanup crew) and helps manage oxidative stress. It also protects against some of the metabolic byproducts that hurt your bladder lining. If you're using K more than once in a while, NAC should be part of your stack. It's cheap, safe, and may save you from some deeply un-fun medical bills.

Nutritional Support

While ketamine doesn't deplete you the way many drugs do, it can still leave you feeling raw. A few supplements that may help:

NAC (N-Acetylcysteine) (600 mg): A precursor to *glutathione*, your body's primary antioxidant. The leading theory for ketamine's damage to the bladder lining points to oxidative stress and direct toxicity from the drug and its breakdown products, which is why NAC is the supplement most regular users reach for first. Essential for Ketamine use.

Magnesium glycinate (200 mg): Magnesium is your body's own NMDA receptor blocker, sitting in the same channel ketamine acts on, which makes it a natural fit here. In practice it eases muscle tension, settles the nervous system, and can take the edge off the wrung-out, jittery feeling that sometimes follows a session. The glycinate form absorbs well and is gentle on the gut, unlike cheaper forms that will send you looking for a bathroom.

Choline (Alpha-GPC or CDP-choline) (300 to 600 mg): Choline is the raw material your brain uses to make acetylcholine, the neurotransmitter behind memory, focus, and clear thinking. Ketamine can leave you foggy and slow afterward, and topping up choline gives your brain the building block it needs to come back online faster. Alpha-GPC and CDP-choline (citicoline) both reach the brain well; CDP-choline tends to feel smoother, Alpha-GPC a little stronger.

Omega-3s (1 to 2 g EPA/DHA): EPA and DHA are the structural fats your neurons' membranes are built from, and they're strongly anti-inflammatory. Since inflammation looks like part of ketamine's toll on both the brain and the bladder, omega-3s support recovery on both fronts and help keep your neurons resilient over the long haul.

B-complex: The B vitamins are the workhorses behind energy metabolism and neurotransmitter production, and they get drawn down whenever your nervous system is under load. A solid B-complex keeps that underlying machinery supplied. This one is general maintenance rather than strictly ketamine-specific, but it's cheap insurance.

Other Considerations

Outside clinics, ketamine remains popular among psychonauts, artists, and partygoers. It offers a fast escape, a pocket-sized dissociative journey, but it carries real risks, especially with frequent or compulsive use.

You don't walk on ketamine; you swim through gravity. Falls are common. Furniture is less friendly than you remember. Use it sitting or lying down, in a space where you won't knock your teeth out trying to stand up and remember what your legs do. Ideally, soft floors, dim lighting, and no sharp corners. K, as they say, is for "*Kouch.*"

Don't go solo. Even if you're chasing internal silence, have someone present. A sitter, a friend, even just a voice in the next room. Dissociation can be healing, but it can also turn disorienting fast, especially if you're new to it or dosing high.

Avoid mixing K with other downers like alcohol, benzos, or opioids. Ketamine already messes with balance, breathing, and blood pressure. Stack those risks and you're playing Jenga with your central nervous system.

Finally, take some time to reflect. Don't just crawl out of the hole and move on. Integration matters. Write about what you saw. Talk it out. Sit in stillness the next morning with a glass of water and let the echoes settle. Some of what ketamine shows you doesn't arrive in words, but it leaves deep impressions. Pay attention to those.

Ketamine, used with intention, can clear emotional static and reset the nervous system like few things can. Used casually and often, it can erode the very self you're trying to recover. That comes from people who have done it both ways.

Walk with care. And if you find something in the silence, pay attention.

DMT/Ayahuasca

"Would you tell me, please, which way I ought to go from here?"

Few molecules obliterate the sense of self as completely or as rapidly as DMT. Inhaled or infused, it launches consciousness into a space that is neither dream nor metaphor. Users speak of fractal palaces, geometric entities, and direct communication with "the intelligence behind reality." It lasts only minutes, but can feel like lifetimes.

When taken orally in the traditional Amazonian brew *ayahuasca*, the experience is slower, longer, and deeply introspective, filled with profound visions, and often, purgation. Both are tools of disintegration, insight, and sometimes, healing, but they are not safe by default. They are not what many would consider "recreational drugs."

DMT is not casual. What it demands goes beyond preparation, and lies closer to reverence than caution.

Historical and Cultural Context

DMT may be just coming into the public eye, but it's not new. Its roots stretch back through centuries of indigenous plant use in the Amazon, where it is consumed not as a pure molecule but in the form of *ayahuasca*, a brew made from two key ingredients: the *Banisteriopsis caapi* vine and the *Psychotria viridis* leaf. This combination represents a pharmacological sophistication developed over generations, arrived at through observation, experimentation, and a relationship with the plant world that Western science is only beginning to take seriously.

The vine contains harmala alkaloids (natural MAO inhibitors) that allow the DMT in the leaf to become orally active. On its own, DMT is rapidly broken down in the gut before it can cross into the brain. With MAOIs in play, that breakdown is blocked, and the trip unfolds over hours rather than minutes. How indigenous peoples discovered this specific combination, out of the tens of thousands of plants in the Amazon basin, remains one of the more compelling mysteries in ethnobotany. The common answer, offered without irony, is that the plants told them.

In these indigenous contexts, ayahuasca is more than a chemical. It is medicine, teacher, spirit, and ceremony. It is used for healing physical and psychological illness, for resolving conflict within communities, and for maintaining relationships with ancestors and unseen worlds. The shaman, or *ayahuasquero*, is a trained specialist who may spend years or decades learning to navigate these realms and guide others through them. The ceremony itself is tightly structured, involving sung healing chants called *icaros*, strict dietary protocols known as *la dieta*, and a collective intention that shapes the space before anyone drinks a drop. The brew is not taken casually, and the setting is not incidental. It is considered part of the medicine.

This framework traveled slowly and unevenly into Western awareness. In the mid-20th century, Hungarian chemist Stephen Szára first administered DMT to human subjects, documenting its psychedelic effects in a clinical context. His work confirmed what indigenous traditions had always known: that this molecule does something categorically different from other psychoactives. The scientific community filed it away. Decades passed.

The profile of DMT changed significantly in the 1990s with Rick Strassman's federally approved research at the University of New Mexico, the first sanctioned psychedelic research in the United States in over two decades. Strassman administered intravenous DMT to volunteers in a controlled clinical setting and documented their reports systematically. What emerged was strange by any scientific standard. Subjects consistently described contact with autonomous entities, complex geometric architectures, and experiences of profound significance that resisted easy categorization. His 2000 book *DMT: The Spirit Molecule* brought these findings to a general audience and seeded a generation of renewed interest in psychedelic research and consciousness studies.

Outside the lab, the culture around DMT was already evolving on its own terms. Freebase DMT, typically vaporized or smoked, became more widely available in underground communities throughout the late 20th and early 21st centuries. Its cousin 5-MeO-DMT, found in the secretions of the Sonoran Desert toad and in several plant species, began attracting its own following, known for producing a trip even more total and ego-dissolving than DMT itself. Both compounds are now encountered across a wide range of contexts: solo experimentation, small ceremonial groups, psychedelic-assisted therapy frameworks, and hybrid ceremonies that blend indigenous structure with contemporary therapeutic approaches.

That diversity of context matters. Ayahuasca consumed in a multi-day ceremony in the Peruvian jungle, under the guidance of a lineage-trained *curandero*, is a fundamentally different event than vaporized freebase DMT taken alone in an apartment. The molecule may be the same. The container is not. Harm reduction in this space requires understanding not just the pharmacology but the architecture of the experience, and how much that architecture depends on preparation, support, and intention.

Mechanism of Action: How DMT Does What It Does

DMT is fast, intense, and almost surgical in its precision. It hits like a detonation. To understand how it does that, let's start with what DMT actually is.

DMT stands for *N,N-Dimethyltryptamine,* a molecule that's found in many plants and animals, including, in trace amounts, the human body. It's part of a family called *tryptamines,* which also includes serotonin, melatonin, and psilocin, the active compound in psychedelic mushrooms. Structurally, DMT looks a lot like serotonin. That similarity is key to how it interacts with the brain.

Inside your brain, serotonin acts as a kind of emotional and perceptual regulator. It helps manage mood, sensory input, and the way your brain processes the world. DMT fits into some of the same receptors that serotonin uses. When DMT locks into that receptor, it sends a powerful signal that reshapes how your brain interprets reality.

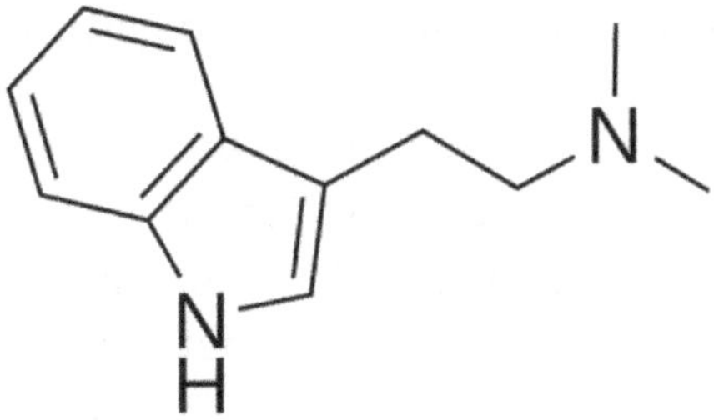

N,N-Dimethyltryptamine, DMT

DMT activates multiple parts of the serotonin system simultaneously. Signals begin to crisscross in unusual ways. Sensory information gets rerouted. Your sense of time, body, language, and memory all begin to unravel and reassemble into something unfamiliar. The visual effects get all the attention, but what's actually happening is a total shift in how your brain organizes consciousness.

When smoked or vaporized, DMT crosses the blood-brain barrier almost immediately. Effects start within 30 seconds, peak around 2 to 3 minutes, and resolve within 10 to 20 minutes. The speed and intensity come from how efficiently the molecule moves into your brain and how quickly it is broken down by enzymes in your liver, particularly *monoamine oxidase*, or *MAO*.

That enzyme is also what prevents DMT from working when taken orally. If you simply swallow DMT, your body neutralizes it before it can enter the bloodstream. That's where *ayahuasca* comes in. The traditional Amazonian brew combines DMT-containing plants with *MAOIs* (MAO inhibitors) from the *Banisteriopsis caapi* vine. These inhibitors temporarily block the enzymes that would destroy DMT, allowing the molecule to enter the bloodstream through the digestive system.

This change in route makes a huge difference. While smoked DMT is fast and overwhelming, ayahuasca is slower, deeper, and longer-lasting. Onset takes 20 to 60 minutes. The peak can stretch over several hours. During that time, the trip tends to unfold in emotional, visual, physical, and psychological waves.

There's also some evidence that DMT interacts with another receptor system called *sigma-1*, which is involved in cellular stress response and possibly in near-death experiences or altered states of identity. This may explain why some users report emotional encounters with entities or sensations of traveling through non-ordinary space.

The truth is, science is still catching up. We know that DMT binds strongly to serotonin receptors. We know it creates dense, synchronized brain activity, and we know that the patterns it produces look a lot like those seen in dreaming or deep meditation, but we still don't fully understand how it produces such specific, repeatable effects, or why people from different backgrounds describe similar spaces, similar shapes, or even similar beings.

What we do know is that DMT operates differently from anything else in this book. It transforms the way your brain models reality. No pleasure center activation, no dulling, no numbing. What you experience is less a distortion of the world than the sudden presentation of another one. Whether that world is internal, external, or both is beyond the scope of science.

Some describe it as a dream with your eyes open. Others call it a glimpse of the divine. The label matters less than the mechanism. Whatever the label, DMT changes the process that assembles perception itself.

The Experience

When smoked, DMT hits in under 30 seconds. There is no ramp-up. The visual field fractures, warps, and explodes into geometry. The body often disappears from awareness. Language stops functioning. For 5 to 15 minutes, users may undergo *total ego dissolution*: a state beyond thought, identity, or context.

Ayahuasca, by contrast, comes on slowly, over 20 to 60 minutes, and can last 4 to 6 hours. The effects are layered: body, emotion, vision, meaning. Purgative effects (vomiting, diarrhea, tears, sweating) are common and traditionally understood as part of the cleansing process. Emotionally, it may bring repressed trauma to the surface, challenge one's worldview, or catalyze life-changing insight.

Note that it's not always pleasant. Many report difficult journeys: terror, physical discomfort, reliving past pain. The line between therapeutic and traumatic can blur without preparation and integration.

Guidelines for Safe Use

DMT asks more of you than caution. It asks for intentionality. The risk with this molecule isn't overdose. It's coming apart at the seams.

Start with your environment. Physical safety, emotional readiness, and a supportive space are baseline requirements. This is not an impulse decision. If you're working with ayahuasca, find a guide or sitter you trust, ideally someone who has done the ceremony before and knows what to expect. If it's your first time, having someone who can hold the space while you go deep isn't a luxury. It's a necessity. Even with smoked DMT, don't do it alone.

If you're smoking DMT, stay seated, and have a place you can quickly lay down. The onset is *very* rapid and can take you by surprise.

Know your source. Plant-based ayahuasca brews vary wildly in potency. DMT extract should be pure, and freebase DMT should be vaporized, not burned, to avoid decomposition. If you can't verify what you have, reconsider whether this is the right time.

Be aware of MAOI interactions. Harmala alkaloids interact with many medications, especially SSRIs, stimulants, and other antidepressants. These interactions can be fatal. See the interaction section below.

Prepare psychologically. Journal beforehand. Set an intention. Don't expect bliss. Expect truth, and be open to whatever form it takes.

Nutritional and Neurochemical Aftercare

DMT doesn't deplete serotonin the way MDMA does, but it activates the brain intensely and often leaves you physically and emotionally raw. Give your body something to work with afterward.

These are a good starting place:

- **Choline (Alpha-GPC or CDP-choline) - 300-600 mg**
- **Omega-3s - 1-2 g EPA/DHA**
- **Magnesium glycinate - 200 mg**

Coconut water or electrolyte-rich fluids help restore balance after purging, especially with ayahuasca. Choline and omega-3s support membrane repair and neuroplasticity. Magnesium helps reduce body tension and support deep sleep. Protein-rich meals (eggs, fish, legumes) give your body the building blocks for neurotransmitter recovery, but stay away from processed meats and aged cheeses if you've had MAOIs. Gentle movement like walking, stretching, or yoga helps your body catch up with what your mind just went through.

Don't skip integration. The visions fade, but the meaning often takes days or weeks to digest. Talk with a trusted friend, therapist, or guide. Journal. Revisit your intentions. DMT can show you a lot in a very short window. Making sense of it is slower work, and it matters more than the experience itself.

What Not to Mix With DMT or Ayahuasca

Mixing DMT with other substances is generally **strongly discouraged**, not because of interaction toxicity in the classic sense (as with cocaine or alcohol), but because of **psychiatric destabilization** and rare but dangerous metabolic consequences.

Avoid combining with:

- **MDMA/MDA, SSRIs, SNRIs, TCAs**: These can interact dangerously with MAOIs in ayahuasca, potentially leading to *serotonin syndrome*: a life-threatening condition. Do not take Ayahuasca if you are on any kind of antidepressants, full stop.
- **Stimulants**: Combining harmala alkaloids with amphetamines, Adderall, cocaine, or even high-dose caffeine can result in *hypertensive crisis* (dangerously high blood pressure), stroke, or seizure.
- **Fermented Foods, Beer, and Cheese:** Fermented foods like aged cheese, beer, and some smoked meats contain large amounts of the chemical *tyramine*. Normally, tyramine is broken down by the same pathway as DMT, but when combined with an MAOI, like in ayahuasca, it can cause spikes in blood pressure. Give it a day or two before you reintroduce fermented foods.
- **5-MeO-DMT**: While pharmacologically distinct, combining it with DMT or taking both in close succession can lead to destabilizing effects and psychological overwhelm.

- **Cannabis**: Some find it softens the edges of DMT, others report it distorts and intensifies visuals in unpredictable ways. Not recommended unless well-known to the user.
- **Benzodiazepines or Alcohol**: May blunt the experience but increase risk of nausea, sedation, and disorientation, especially in ayahuasca ceremonies where purging is expected.

Ayahuasca traditions often recommend a *"dieta,"* a restricted diet leading up to the ceremony that avoids red meat, processed foods, alcohol, sex, and certain medications. This isn't just superstition. Many of these guidelines help reduce MAOI interaction risk and prepare the body for deeper work.

Closing Note: The Mirror Behind the Mirror

DMT doesn't ease you in. It doesn't offer comfort or closure or a gentle reframe. It drops you into something that has no name and no edge, and when it's over, you come back holding nothing but the memory of having been there. Ayahuasca takes longer and goes deeper, working through your history with a patience and thoroughness that feels less like exploration and more like excavation.

Every culture that has worked seriously with these substances has surrounded them with structure: fasting, ceremony, intention, a guide who knows the terrain, days of preparation and days of recovery. That structure exists for a reason. People who skip it tend to find out why.

Respect the weight of what you're walking into. What you encounter there will be real. Make sure you are too.

Conclusion – Psychedelics and the Human Story

"Begin at the beginning," the King said, gravely, "and go on till you come to the end: then stop."

Humans have been using psychoactive substances for as long as we have records. In ceremonies and healing work, in moments of curiosity and desperation, we've turned to plants and molecules looking for meaning. Psilocybin mushrooms, vine brews, cactus buttons, fermented grains. The specific substance mattered less than the impulse: to step outside the ordinary machinery of the mind and see what's there.

That impulse shows up in every culture. Not because we're broken or lost, but because altering consciousness is something humans do. The question was never whether people would engage with these substances. It was always how deeply they would go.

This book has tried to give you an honest answer to that question. It started with the brain itself: the neurotransmitters and electrical signals that produce thought, emotion, identity. Then it traced what happens when you introduce MDMA, LSD, cannabis, ketamine, psilocybin, and DMT into that system. How they amplify certain signals, suppress others, open doors you didn't know were there. Sometimes those doors lead somewhere useful. Sometimes they don't.

One idea has stayed constant through all of it. These substances are powerful because they amplify your brain's own function. They show you what is already inside, stripped of the filters you've spent a lifetime building. That can be revelatory. It can also be destabilizing. Often it is both at once.

That power demands respect, and respect here means understanding. The principles of harm reduction in this book are practices drawn from biology, pharmacology, experience, and hard-earned wisdom, including wisdom from communities that have been working with these substances for centuries, long before Western science arrived to take notes.

Harm reduction was never only about avoiding the worst outcomes. That's the floor. The deeper invitation is to **pay attention**. Pay attention to your body and what it's telling you before, during, and after. To your mind, including the parts of it you'd rather not look at. To your setting, your intention, and the people around you. To the source and content of what you're taking, because that still matters even when you think you know what you have. And especially to what comes up. Drugs may open the door, but what you do in the room is entirely up to you.

For many people, psychedelic exploration feels less like a departure from ordinary life than a return to something they didn't know they'd lost. These states are latent possibilities of the human brain, accessible under specific conditions, pointing toward the fact that the ordinary waking mind is one mode among many. That ego is one identity among several. That the story you tell about yourself can be rewritten.

This is why these substances have been used across history in rites of passage, in mourning, in healing, and in communion. They confront us with our own material, with grief we've buried, fears we've organized our lives around, patterns we inherited and never questioned. And sometimes, under the right conditions, with the right support, they create enough distance from that material that we can actually look at it. Sometimes, people rewrite things they had assumed were permanent.

The renewed scientific interest in psilocybin for depression, MDMA for PTSD, and ketamine for treatment-resistant conditions represents a rebalancing. For decades, research into these compounds was stifled by politics and stigma, by scheduling decisions that were never really about science. Now, slowly, we are catching up. The early results are striking enough that researchers who spent careers being cautious about making strong claims are starting to make them. That deserves attention.

At the same time, the fact that something works in a clinical trial says little about how it works in other contexts. The therapeutic framing, the preparation, the integration support, and the relationship with the guide, are all part of the treatment. Extracting the molecule and ignoring the container has never worked as well as the full architecture. That lesson applies whether you are sitting in a research clinic or in your living room with friends.

As laws shift and new generations engage with these substances, the responsibility on all of us grows. To stay safe, yes, but also to stay grounded in what these substances have actually been used for, across cultures and centuries, before anyone wrote a trip report. To ask what this is showing you, what it might be asking of you, and whether you are actually doing anything with what you learn. Integration isn't a buzzword. It's the whole thing. The trip without integration is just a trip. What you carry out of the room and into your actual life is what determines whether it mattered.

That is where the future of this conversation lives. In better questions. Questions about the nature of consciousness and suffering and meaning and connection. Psychedelics don't answer those questions. They make it harder to pretend the questions don't matter.

If you have read this book to the end, you are already asking them. Keep asking. Stay curious. Take care of your mind, your body, and your people. Let experience become understanding, and let understanding become something you actually act on.

You were never static. Every choice, every molecule, every moment you actually paid attention to has shaped what you are now. The ones you sit with long enough to integrate will shape what comes next.

Listen closely. Respect the chemistry. Respect yourself.

And test your frickin' drugs. ♥

Appendices

The appendices that follow are here for one purpose: to give you practical tools you can use in the real world. The chapters before this point explain what substances do, how they alter perception, and how you can navigate those experiences with care. The appendices shift from exploration to application. This is the part of the book you can return to before a long night out or while preparing a safer space for friends who are heading into something unfamiliar. These sections are meant to be handled, bookmarked, and used.

Although these appendices are technical, they were written with the same philosophy that guides the rest of the book. The goal is not to encourage use. The goal is to reduce harm. If you choose to engage with substances, you deserve clear information and dependable instructions rather than guesswork or hearsay. Honest preparation is part of respecting the journey.

Think of this final section as a toolkit. You may not need everything in it now, but it should be available if you ever do. Learning these skills builds confidence, and wisdom protects you far more effectively than fear. Whether you are caring for yourself or for others, the appendices offer the practical knowledge required to make informed, grounded decisions. They are the quiet work that supports the rest of the journey.

How to Use a Fentanyl Test Strip

Fentanyl is a synthetic opioid that can kill in micrograms. It shows up where it doesn't belong: in pressed pills, cocaine, ketamine, MDMA, methamphetamine, even fake Xanax bars. You can't see it. You can't taste it. You can only test for it.

Fentanyl test strips are cheap, fast, and save lives. Originally designed for urine, they're now one of the most effective off-label harm reduction tools available. Testing takes less than 2 minutes. If you use drugs, you should test. Every batch. Every time.

Where to Get Them

Fentanyl test strips are legal in almost all U.S. states and can be ordered discreetly online. Reputable sources include:

- DanceSafe.org
- BunkPolice.com
- Dosetest.com

Some community health organizations, needle exchanges, and music festivals distribute them for free. Check local harm reduction programs or syringe access centers.

Check the dates on your test strips, and make sure you store them in a cool, dry place. Leave them in the package until you're ready to use them. While some sources suggest they may remain functional for some time beyond the expiration date (up to 5 years), their trustworthiness decreases as they get older.

What You'll Need

- A fentanyl test strip (from a trusted source)
- A small amount of the substance (a full dose or a few scrapings)
- Clean water (at least 1 mL; distilled or bottled preferred)
- A clean container (spoon, bottle cap, shot glass), or the baggie your drugs came in

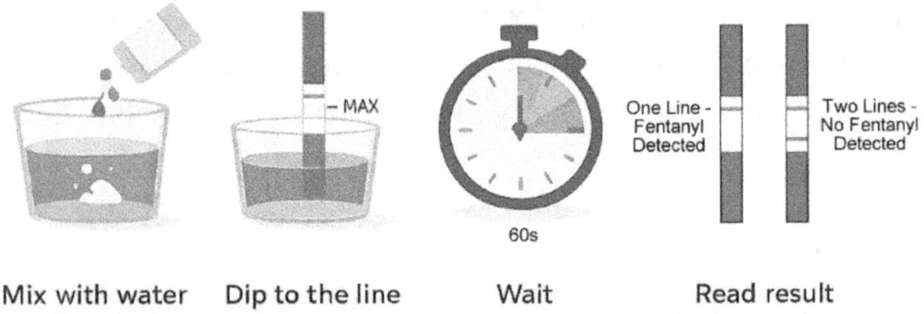

Step-by-Step Instructions

Option 1: Standard Method

1. Prepare a sample. Put a very small amount of your drug (no more than a match-head size) into a clean container. Add about 1 mL (¼ teaspoon) of water and stir until fully dissolved. Too much powder can overwhelm the test and cause a false negative, especially with stimulants.
2. Dip the strip. Hold the strip by the solid end. Dip the wavy end into the liquid for 15 seconds. Don't submerge past the max-fill line.
3. Lay the strip flat on a clean surface. Wait 60 seconds.
4. Read results, as below.

Option 2: Bag Test

If you've already used the dose, don't have a container handy, or only have trace residue left:

1. Add about 1 mL (¼ teaspoon) of water directly into the empty bag the drugs came in.
2. Swirl it around to dissolve any leftover material.
3. Dip the test strip in the same way: 15 seconds, wavy end down.
4. Wait 60 seconds and read results as below.

Reading the Results

- One line = Positive (fentanyl detected)
- Two lines = Negative (no fentanyl detected)
- No lines = Invalid (retest)

Note: Even a faint second line still means negative. Faint ≠ positive.

Important Reminders

- Always test powders and pressed pills, even if they came from someone you know.
- Fentanyl test strips may not detect every novel synthetic opioid (such as nitazenes), but they're still your best defense.
- Never use alone. Carry naloxone. Even if you don't plan to use opioids, fentanyl might be present anyway.
- A negative result is not a safety guarantee, but a positive result is a flashing red light.
- There's no trophy for trust. Test everything. Every time.

How to Use a Milligram Scale

You wouldn't eyeball your insulin. You wouldn't guess how much LSD is on a tab. Don't guess with powders either. A scale is a form of self-respect.

Most psychoactive compounds are active in the **milligram** range. That's thousandths of a gram. A standard capsule holds 500-1000 mg of material. The difference between a helpful dose and a hospital visit can be 10 mg or less.

If you're working with MDMA, ketamine, 2C compounds, synthetic tryptamines, or almost anything that comes as a powder or crystal, you need a scale. Not a kitchen scale. Not your friend's "eyeball method." Not a finger-dip. A real digital scale that measures to the **milligram (mg)** (0.001 g).

What to Look for in a Scale

Look for a **0.001 g digital scale** (also called a *milligram scale*) with:

- **Calibration weights** (50 g or 100 g preferred)
- **Tare function** (to zero out your container)
- **Backlit screen** (for nighttime prep)
- **Battery power** (for flexibility)

Reliable brands include Gemini-20, AWS, Smart Weigh, and Weigh Gram. Expect to spend $20-50. Look for "jewelry" or "gunpowder" scales.

If you find yourself using your scale a lot, it might also be worthwhile to pick up some *weighing papers.* These are specially coated non-stick sheets of paper, and meant to be disposable. It makes working with powders easier (especially in the milligram range).

Scales are most accurate in the middle of their range. If you're weighing 30-100 mg doses, place a calibration weight on the scale and subtract it from your total. (Example: I add a 10 g weight to the scale. If I'm trying to measure out 75 mg, I would want the scale to read 10.075 g, that is, 10 grams plus 75 milligrams.)

How to Use It

1. **Place the scale on a flat, stable surface.**
   No couches or beds. No vibrating speakers. No fans or breezes.
2. **Calibrate it using the included weight.**
   Follow the manual. Calibration ensures accuracy, and accuracy is everything.
3. **Tare your container.**
   Place your capsule, weigh paper, or spoon on the scale. Hit **TARE** to zero it out.
4. **Add a calibration weight to the scale.**
   Read the scale. Write down the weight of the calibration weight, if you need to.
5. **Add the substance slowly.**
   Use a micro-scoop or folded paper. Tap lightly. Go slow. Small changes make big differences.
6. **Verify the dose.**
   Read the scale. Subtract the calibration weight, and that's your final number. Overshot it? Remove a little. Too light? Add more. Be exact. Be patient.

Best Practices

- Turn on the scale, and let it warm up for 30-60 seconds before using it, especially for low doses (less than 100 mg).
- Always confirm you're reading in grams (g), not ounces (oz) or grains (gn).
- Double-check your decimal places. 0.001 g is 1 mg. 0.010 g is 10 mg. 0.100 g is 100 mg. Read carefully.
- Wear latex/nitrile gloves, if possible. Dispose of them afterwards. Avoid touching powders with your fingers. Oil, moisture, and drug absorption are all risks.
- Keep the scale clean and dry. Store it in a hard-shell case. Try to avoid touching the weighing platform with your bare hands. Fingerprints weigh milligrams too!
- Recalibrate monthly or after transport. Follow the instructions in the manual.
- Practice with salt or sugar if you're new to using the scale. Get the feel for milligrams before your life depends on it.

Imperial-to-Metric Conversion

If you need to go from ounces to grams, here's a quick reference, in fractions and decimals. One pound (lb.) is 16 ounces.

Ounces (Oz)		Grams (g)
1/10 oz	0.100 oz	2.835 g
1/8 oz	0.125 oz	3.544 g
1/4 oz	0.250 oz	7.087 g
1/3 oz	0.333 oz	9.450 g
1/2 oz	0.500 oz	14.175 g
2/3 oz	0.667 oz	18.900 g
3/4 oz	0.750 oz	21.263 g
1 oz	1.000 oz	28.350 g

How to Use a Drug Test Kit

If you choose to use psychoactive substances, the most important step you can take toward safety is to verify what you're taking. A reagent test kit is a simple chemical tool that can identify many adulterants or confirm the presence of a desired compound. No test is perfect, but a few drops of liquid can mean the difference between a known risk and a fatal mistake.

Testing kits can be obtained online from DanceSafe.org, BunkPolice.com, or many other reputable sellers or community organizations. These kits typically consist of one or more dropper bottles full of chemical reagents. There are different tests for different substances. Get a spot kit with many different reagents, even if you don't expect to need them all. Sometimes, they'll catch things you didn't expect!

Treat these kits with respect! They contain toxic and corrosive chemicals, and they might just save your life. Store them in the freezer and they should last about a year.

For testing, you'll need:

- A test kit
- A sample to be tested (the size of a pin head)
- A white ceramic/glass plate
- A well-lit room
- Baking soda (for cleanup)

1. Prepare the surface: Work on a clean, well-lit, non-porous surface. Avoid metal or plastic that might react with the reagents. White ceramic or glass is ideal, especially if you have a plate you can dedicate to it. Don't use plates that you plan on eating from later. If you can find a ceramic palette (often used for painting), that also works very well.

2. Use minimal sample size: Scrape off a small amount, about a grain of salt or the head of a pin. Make several samples, and place them at least a couple of inches apart on your plate. More is not better, since testing destroys the sample.

3. Apply the reagent: Place a drop directly onto the sample and observe the color change. Do not touch the reagent bottle tip to the powder. Apply only one reagent to each sample. Some reagents (such as Simon's reagent) might involve multiple bottles. Your kit will come with instructions on how exactly to use it. If you have multiple bottles, make sure you put the cap on one bottle of reagent before you open another one! Mixing the caps will ruin your test kits, and is incredibly easy to do in the excitement of testing.

4. Compare colors quickly: Reaction colors change over seconds. Use the manufacturer's color chart immediately to match results. Record the time and color for reference. Note if there was any bubbling, and if there were any intermediate colors. Did it turn purple first, and then black? Or was it green first? Are there multiple colors present? It often happens fast, so be observant! Some tests (such as Ehrlich), may take as long as 5 minutes to complete. Your kit will come with a chart indicating what colors to expect, and in what time.

5. Cross-test. Each reagent (Marquis, Mecke, Mandelin, Ehrlich, Hofmann, etc.) detects different substances. It will take a combination of a few of them to make a positive determination of what is and isn't in your sample. A single reagent can only rule out so much. Use at least two complementary reagents for confirmation. For example, MDMA shows purple with Marquis and green with Mecke. Look at the chart accompanying your test kit, and pick at least two reagents to confirm. Better yet, try them all!

6. Interpret carefully. A positive color match indicates that the substance likely contains that compound, not that it is pure or safe. Think of it this way: Your goal when testing is to verify that nothing unusual is happening; if you have unexpected colors or reactions, *something unusual has happened.* This could signal possible adulteration. Don't try to guess why your sample tested differently; just don't take it. Without lab testing, it's impossible to know if it's due to impurities from manufacture or dangerous analogs.

7. Dispose safely. Neutralize the tested residue with baking soda and rinse with water. Store reagents tightly capped and away from heat or light; in the back of your freezer is ideal.

Getting strange results, and think the kit might be expired, tainted, or no longer potent? Try testing aspirin or sugar as a "control." You can find color charts for these with many test kits or online. If your control substance also tests strangely, you'll know that your test kit is expired. Neutralize your remaining reagents with baking soda and dispose of them.

Remember: reagent testing cannot confirm dose, purity, or safety, but it can prevent tragedy. Knowledge saves lives. Respect the chemistry and yourself. Test every time.

Quick Reference for Harm Reduction Supplements

This is your fast-glance protocol guide. Use it in the field, on the night, or on the comedown couch. Based on the best available evidence, field reports, and years of lived experience.

General Supplement Advice

When you're purchasing vitamins, a couple of things to note:

- When buying magnesium supplements, be sure to check the form. Get Magnesium **Glycinate**, not Magnesium Citrate or Oxide. Magnesium Glycinate is more easily absorbed into your body, whereas the other forms tend to have a strong laxative effect.
- Many supplements come in chewable gummy form. Get vitamins you enjoy taking, and won't bail on in the middle of your experience. Be honest with yourself about those giant pills. Will you want to choke those down mid-trip?
- Check how many pills are in a serving. Sometimes, you need to take more than one to get the advertised dose.
- Check the expiration dates on your supplements. Despite our best wishes, they don't last forever.
- Tired of juggling all those bottles? A pill organizer can make preparation much easier. Look for one with a waterproof lid.

LSD

Timing	Supplements / Strategies
During	Choline (Alpha-GPC or CDP-choline) - 300-600 mg
After	Choline - 300-600 mg Magnesium glycinate - 200 mg Vitamin C - 1000 mg Omega-3s - 1-2 g EPA/DHA B-complex Sleep & Integration work

Psilocybin Mushrooms

Timing	Supplements / Strategies
During	Ginger or peppermint (optional, for tea or nausea) Lemon juice, for "Lemon tek" (optional)
After	Choline (Alpha-GPC or CDP-choline) - 300-600 mg Magnesium glycinate - 200 mg Omega-3s - 1-2 g EPA/DHA Vitamin C - 1000 mg Gentle food & water Space to reflect and integrate

MDMA/MDA

Timing	Supplements / Strategies
6 Hours Before	Magnesium glycinate - 200 mg Na-R-ALA - 100 mg Fresh ginger - ≈3 g
1-3 Hours Before	Magnesium glycinate - 200 mg Na-R-ALA - 100 mg Fresh ginger - ≈3 g N-Acetylcysteine (NAC) - 600 mg Vitamin C - 1000 mg Grape seed extract - 100 mg ALCAR - 500 mg EGCg - 400 mg L-Tyrosine - 500 mg CoQ10 - 100 mg
30 Min Before	Tums or Rolaids - 2 tablets
During (Every 2 Hours)	Magnesium glycinate - 200 mg Na-R-ALA - 100 mg Grape seed extract - 100 mg N-Acetylcysteine (NAC) - 600 mg

MDMA/MDA (continued)

After, Same Night	Melatonin - 10 mg Electrolytes/Sports Drinks (not just water)
Next 2-3 Days, Before Bed	5-HTP - 100-200 mg (*after 12-24 hrs*) EGCg - 400 mg Vitamin B-6 - 25-50 mg Magnesium glycinate - 200 mg Melatonin - 1-5 mg, as needed

Cannabis

Timing	Supplements / Strategies
Before	CBD - 25-50 mg Magnesium glycinate - 200 mg
During (if anxious)	Black pepper - a few peppercorns Fresh lemon peel CBD - 25-50 mg
After (regular/ heavy users)	Omega-3s - 1-2 g EPA/DHA Citicoline (CDP-choline) - 300-600 mg Melatonin - 1-5 mg Magnesium glycinate - 200 mg

Nitrous Oxide

Timing	Supplements / Strategies
Before/ After	Methylcobalamin B-12 - 2000 mcg (sublingual) Methylfolate (5-MTHF) - 400-800 mcg Vitamin B-6 (as P-5-P) - 25-50 mg Riboflavin/B-2 - 100 mg L-Methionine - 500 mg Breathe air between hits Wait between sessions NEVER inhale directly from canister

Cocaine

Timing	Supplements / Strategies
During	Electrolyte fluids Saline nasal rinse or neti pot
1-3 Days After	L-Tyrosine - 500-1000 mg Magnesium glycinate - 200 mg Vitamin C - 1000 mg NAC (N-Acetylcysteine) - 600 mg Rest, food, hydration

DMT/Ayahuasca

Timing	Supplements / Strategies
Before **(Ayahuasca)**	Avoid SSRIs, stimulants, tyramine-rich foods (such as fermented or aged foods)
After	Choline (Alpha-GPC or CDP-choline) - 300-600 mg Omega-3s - 1-2 g EPA/DHA Magnesium glycinate - 200 mg Electrolytes or coconut water Light protein (eggs, fish) Integration (journal, walks, support) Time to decompress

Ketamine

Timing	Supplements / Strategies
After	NAC (N-Acetylcysteine) - 600 mg Magnesium glycinate - 200 mg Choline (Alpha-GPC or CDP-choline) - 300-600 mg Omega-3s - 1-2 g EPA/DHA B-complex Hydration Movement, space between sessions

A note on all of the above: Supplements are not a safety net. They are not permission. They do not make any of these substances safe, and none of what's listed here has been evaluated by the FDA for the purposes described.

You also don't need to take everything listed. Pick what makes sense for your situation, your budget, and your body. Bodies vary. Interactions vary. What works well for one person may not work for another, and no vitamin protocol eliminates the real risks that come with psychoactive substances.

It's also worth knowing that some of the supplements in this book interact with common prescription medications on their own, independent of any recreational substance. NAC, 5-HTP, and magnesium all have documented interactions with certain prescriptions. If you're on medication, check with a doctor or pharmacist before adding anything new to your routine. This is a five-minute conversation that can prevent a bad outcome.

The stacks in this book represent the best available harm reduction thinking at the time of writing, built from community experience and plausible biochemistry. They are a floor, not a ceiling, and they are not a substitute for testing your substances, knowing your doses, and making honest decisions about your own health. When in doubt, talk to a doctor. When you can't, be conservative. The goal is that you're still here to learn from it.

Combo Safety Chart

Mixing drugs always adds risk. This chart offers a quick visual guide to which combinations are less concerning and which are known to be more dangerous.

	Mushrooms	MDMA/MDA	Ketamine	Nitrous	Cocaine	Alcohol	Cannabis
LSD	OK	OK	OK	OK	⚠	▼	⚠
Mushrooms		OK	OK	OK	⚠	▼	⚠
MDMA			OK	OK	⊘	⚠	OK
Ketamine				OK	⚠	⊘	OK
Nitrous					OK	⚠	OK
Cocaine						⊘	⚠
Alcohol							OK

OK	**Low Risk**	Combo not known to be dangerous
▼	**Low Risk/Reduced Effects**	Combo will diminish the effects of both
⚠	**Use Care**	Combo can be unpredictable
⊘	**Dangerous**	Combo is known to be dangerous
☠	**Extreme Danger**	Combo is known to be life threatening

When mixing drugs, keep in mind they'll interact, and start with lower doses of each substance. The following mixtures present elevated risk:

Combination	Risk Level	Reason / Effect
MDMA + Cocaine	🛇 Dangerous	Severe cardiovascular stress
MDMA + MAOIs	💀 Extreme	Serotonin syndrome (potentially fatal)
MDMA + Alcohol	⚠ Use Care	Dehydration, dulling of effects
MDMA + DXM	🛇 Dangerous	Overheating, serotonin toxicity
Cocaine + Alcohol	🛇 Dangerous	Forms *cocaethylene*, highly cardiotoxic
Cocaine + Fentanyl	💀 Extreme	Respiratory arrest, death
Cocaine + Benzos	⚠ Use Care	Masks red flags; false sense of safety
Alcohol + Benzos	💀 Extreme	Both CNS depressants. No safe ratio; can be deadly.

Combination	Risk Level	Reason / Effect
LSD + Antipsychotics	🚫 Dangerous	May dull or worsen the experience unpredictably
Cannabis + Mushrooms/ LSD	⚠ Use Care	May unpredictably intensify experience
Psilocybin + Tobacco	⚠ Use Care	May unpredictably intensify experience
Ketamine + Alcohol	🚫 Dangerous	CNS depression, blackouts
Ayahuasca + SSRIs	💀 Extreme	Serotonin syndrome in ayahuasca settings
Opioids + Anything	💀 Extreme	Severe respiratory depression risk. Carry naloxone.

Remember: This chart is meant as a quick reference guide, and doesn't contain every risky combination. Low risk doesn't mean no risk. Do your own research. Individual needs vary. Start low. You can always take more, but you can't take less. Listen to your body. Hydrate. Sleep. Wait between sessions. And yes, don't forget to test your drugs!

Resources for Further Exploration

Books

PiHKAL and ***TiHKAL*** by Alexander and Ann Shulgin
The field guides for psychedelic chemists and psychonauts. Half chemistry manual, half memoir, these books document the synthesis and subjective effects of hundreds of compounds. Technical and poetic. Legal in most countries.

The Psychedelic Explorer's Guide by James Fadiman
A friendly, practical introduction to safe and intentional psychedelic use. Includes protocol suggestions, trip-sitting advice, and microdosing notes.

Plants of the Gods by Richard Evans Schultes et al.
An incredible look at how psychedelics have evolved through different cultures. A beautiful book, it truly explores the universal human nature of psychedelics.

How to Change Your Mind by Michael Pollan
Narrative journalism meets psychedelic renaissance. Covers psilocybin therapy, LSD research, and the modern return of psychedelics to psychiatry. Less technical, more reflective.

DMT: The Spirit Molecule by Rick Strassman
A foundational work in the modern psychedelic science revival. Focuses on clinical trials with DMT, its pharmacology, and speculative links to consciousness and dreaming.

A Brief History of Drugs by Antonio Escohotado
A great exploration of the impact of drugs throughout history, across cultures, and around the world.

Websites

Erowid.org

The original vault. Trip reports, dosage guides, chemical profiles, and legal information. Crowdsourced but moderated. Check the experience level of contributors.

PsychonautWiki.org

An open-source encyclopedia of psychoactive substances. Focuses on subjective effects, mechanisms of action, safety, and trip reports. More structured than Erowid, with chemistry notes and receptor binding charts.

Tripsit.me

Focused on practical support, combination safety, and live trip-sitting chat. Their combo chart is essential: it shows which drugs are safe, risky, or deadly to mix. Keep this one bookmarked.

DanceSafe.org

Pioneer in festival harm reduction. Offers test kits, safety literature, and science-backed advice. Also includes up-to-date alerts on adulterants and counterfeit pills in circulation.

MAPS.org

The Multidisciplinary Association for Psychedelic Studies. Central hub for clinical research, particularly around MDMA-assisted therapy. Excellent for reading studies, trial outcomes, and political policy updates.

RollSafe.org

Harm reduction site focused on MDMA safety. Great resource covering many aspects of MDMA harm reduction.

BunkPolice.com
Supplier of advanced reagent testing kits and educational tools for substance identification. Also documents deceptive or adulterated products being sold globally.

ZendoProject.org
Psychedelic peer support training and infrastructure. Teaches you how to help someone through a difficult trip without calling an ambulance unless necessary. Grounded, compassionate, trauma-aware.

AlteredStatesBook.com
The website for this book. Includes my blog, my bibliography, and links to the sources and scientific research supporting this book. Visit to order additional copies of the book, download harm-reduction cards, or contact the author.

Addiction & Recovery Resources

If your relationship with substances has shifted from something you choose into something that chooses you, help is available. The resources below are organized by region. Many are free, confidential, and available around the clock. You don't need to have hit rock bottom to reach out. You just need to be willing to pick up the phone, send the text, or visit the website.

Addiction is not the opposite of harm reduction. Recognizing when you need help is one of its highest forms.

Everywhere

SMART Recovery
smartrecovery.org

A science-based alternative to 12-step programs, focused on self-empowerment and cognitive behavioral tools. Free online meetings are available daily and open to participants worldwide.

Alcoholics Anonymous/Narcotics Anonymous
aa.org / na.org

The original peer support networks for substance use, available in nearly every city in the world. Meetings are free, open, and anonymous. Both organizations maintain international directories on their websites.

The Fireside Project
Text or call 62-FIRESIDE (623-473-7433)
firesideproject.org

A psychedelic peer support line for people going through difficult trips or needing help integrating challenging psychedelic experiences. Staffed by trained volunteers who understand the territory.

United States

SAMHSA National Helpline
1-800-662-4357 (free, confidential, 24/7)
samhsa.gov/find-help/national-helpline

Referrals to local treatment, support groups, and community organizations. Available in English and Spanish. They don't ask for insurance, and they don't report you to anyone.

988 Suicide & Crisis Lifeline
Call or text 988
988lifeline.org

For anyone in crisis, struggling with suicidal thoughts, or experiencing a substance-related emergency. Available 24/7.

Never Use Alone
1-800-484-3731
neverusealone.com

A harm reduction lifeline for people who use drugs alone. An operator stays on the line while you use, and calls emergency services if you stop responding. No judgment. No lectures. Just someone making sure you survive.

Canada

Canadian Centre on Substance Use and Addiction (CCSA)

ccsa.ca

Canada's national authority on substance use. Their website links to provincial and territorial helplines, treatment directories, and educational resources across all provinces. A good first stop if you're not sure where to begin.

Provincial Addiction Helplines

Toll-free & confidential. Each province operates its own addiction helpline. Some of the key ones:

- Alberta (1-866-332-2322)
- British Columbia (1-800-663-1441)
- Ontario/ConnexOntario (1-866-531-2600)
- Quebec/Drugs: Help and Referral (1-800-265-2626)
- For others, dial 811 or visit the CCSA website

988 Suicide Crisis Helpline (Canada)

Call or text 988

Canada launched its own 988 service in 2023. Available 24/7 for anyone in crisis, including substance-related emergencies.

Centre for Addiction and Mental Health (CAMH)

camh.ca

Canada's largest mental health and addiction teaching hospital, based in Toronto. Their website provides assessment tools, treatment information, and resources.

United Kingdom

Talk to FRANK
0300 123 6600 (Available 24/7) or text 82111
talktofrank.com

The UK's national drug information and advice service. Free, confidential, and available in over 120 languages. Provides drug information, local service directories, and support for users, families, and friends.

NHS Drug Addiction Support
nhs.uk/live-well/addiction-support

The NHS provides free treatment for drug addiction, including counseling, medication-assisted treatment, and residential programs. You can refer yourself directly to local drug treatment services without going through your GP.

European Union

European Union Drugs Agency (EUDA)
euda.europa.eu

Formerly the EMCDDA, this is the EU's central resource for drug information, country-by-country treatment data, and harm reduction policy. Their website links to national drug strategies, treatment availability, and drug checking services across all EU member states.

Drug treatment services in most EU countries are publicly funded and available without charge. Contact your national health service or search the EUDA website for treatment options in your country.

LGBTQ+ and Underserved Communities

Substance use affects everyone, but not everyone has equal access to support that understands their experience. LGBTQ+ individuals face substance use disorders at roughly two to three times the rate of the general population, often rooted in stigma, discrimination, and the long cultural role of bars and nightlife as the only safe social spaces. Indigenous and BIPOC communities carry the additional weight of systemic inequity, intergenerational trauma, and a long history of being policed rather than helped. The resources below are built by and for these communities.

The Trevor Project
Call 1-866-488-7386 or text START to 678-678
thetrevorproject.org

Crisis intervention and suicide prevention for LGBTQ+ young people. Trained counselors available 24/7 by phone, text, and online chat. If you're a young queer person in crisis, this is the call to make.

LGBT National Help Center
1-888-843-4564
lgbthotline.org

Free, confidential peer support for LGBTQ+ people of all ages. Also operates a youth line (1-800-246-7743) and a senior line (1-888-234-7243). Available Monday through Friday, with limited weekend hours.

Thrive Lifeline
Text THRIVE to 313-662-8209

A text-based crisis line staffed by people with lived experience in marginalized communities, including LGBTQ+, BIPOC, and disabled individuals. When mainstream crisis lines don't feel safe, this one might.

GLMA Provider Directory
glma.org

The GLMA (Health Professionals Advancing LGBTQ+ Equality) maintains a directory of LGBTQ+-affirming healthcare providers, including addiction specialists. Useful for finding treatment where you won't have to explain or defend your identity.

White Bison / Wellbriety Movement
whitebison.org

A grassroots nonprofit providing culturally grounded recovery support for Native American and Alaska Native communities. The Wellbriety Movement offers mutual aid meetings, addiction prevention programs, and training rooted in Indigenous healing traditions. Active across the US and Canada.

National Harm Reduction Coalition: Native Harm Reduction Toolkit
harmreduction.org/native-toolkit

A living resource developed to support Indigenous communities in building harm reduction services grounded in decolonized, culturally responsive practice. Includes guides, fact sheets, and frameworks for Indigenizing harm reduction.

One Sky Center
oneskycenter.org

Works to improve substance use prevention and treatment for Indigenous peoples through mentorship, education, and connections to culturally appropriate care.

You don't need to be in crisis to use these resources. You don't need to have a diagnosis, a plan, or the right words. You just need to reach out. That's enough.

Glossary of Terms

5-HT/5-HT2A Receptor

5-HT is serotonin. The 5-HT2A receptor is the serotonin subtype that most classic psychedelics latch onto to alter perception and thought.

5-HTP (5-Hydroxytryptophan)

An amino acid and serotonin precursor, taken during recovery to rebuild serotonin. Pair it with EGCg, and never with active serotonergic drugs.

Acetyl-L-Carnitine (ALCAR)

A supplement that supports mitochondrial energy and mental clarity. Common in harm reduction stacks for recovering from MDMA and stimulant use.

Acetylcholine

A neurotransmitter behind learning, memory, attention, and muscle control. You can support it with nutrients like choline and Alpha-GPC.

Adulterant

Something added to a drug to fake its effects, add weight, or stretch supply. Ranges from harmless caffeine to deadly fentanyl.

Agonist

A substance that binds a receptor and switches it on, imitating the natural neurotransmitter that normally does the job.

Alpha-GPC

A highly absorbable form of choline that crosses the blood-brain barrier. Used to support memory, focus, and acetylcholine production.

Antagonist

A substance that binds a receptor and blocks it, keeping other compounds from switching it on. The opposite of an agonist.

Ayahuasca

A traditional Amazonian brew that pairs DMT-containing plants with MAOIs so the DMT works orally. Used for spiritual and healing work.

Benzodiazepines

A class of sedatives prescribed for anxiety and sleep. Safe enough alone, but dangerous stacked with other depressants like alcohol or opioids.

Blood-Brain Barrier

A tight cellular filter around the brain's blood vessels. It blocks most substances, so any drug that alters consciousness has to cross it.

Body Load

The heavy physical side of a psychedelic: nausea, muscle tension, or fatigue. How rough it feels depends on dose and setting.

Choline

A nutrient the body uses to make acetylcholine, supporting memory and thinking. Often taken to recover after psychedelic or dissociative use.

Cocaethylene

A toxic compound your liver builds when you mix cocaine and alcohol. Harder on the heart than either drug on its own.

Curandera/Curandero

A traditional healer in Indigenous Latin American cultures, trained in plant medicines and ceremony for physical and spiritual healing.

CYP450/CYP2D6/CYP1A2

Liver enzyme families that break down drugs. Ginger, grapefruit, and some medications shift their activity, and genetics make everyone metabolize a little differently.

Default Mode Network (DMN)

The brain network running your self-talk and mind-wandering. Psychedelics quiet it down, which is part of how they loosen fixed patterns of thought.

Dissociation

A state where your sense of body, time, or self goes distorted or disconnected. Common with ketamine and high-dose nitrous oxide.

Dopamine

The neurotransmitter of reward, motivation, and pleasure. Cocaine and methamphetamine flood it, which is a big part of why they hook people.

EGCg (Epigallocatechin gallate)

A strong green-tea antioxidant. Paired with 5-HTP to help serotonin synthesis and blunt side effects, and it mildly inhibits MAO.

Ego Dissolution

The temporary loss of your sense of self and where you end and everything else begins. Often brought on by LSD or psilocybin.

Empathogen

A drug that opens up emotional warmth and a sense of connection. Overlaps with entactogen, and MDMA is the classic example.

First-Pass Metabolism

The liver's first crack at an oral drug, breaking down a chunk of it before it ever reaches your bloodstream.

Glutamate/NMDA Receptor

Glutamate is the brain's main excitatory signal. The NMDA receptor is one of its docking points, and where dissociatives like ketamine act.

Harm Reduction

The practice of making drug use safer rather than just forbidding it, through education, testing, and planning ahead.

Homeostasis

The body's drive to keep its systems in balance, holding things like neurotransmitters, temperature, and blood pressure inside a workable range.

Homocysteine

An amino acid that builds up when B-12 or folate runs low. Measuring it catches a functional B-12 deficiency earlier than standard panels.

Hyponatremia

Dangerously low blood sodium from drinking too much plain water. A real MDMA risk when people overhydrate without replacing electrolytes.

Integration

The work of making sense of a psychedelic experience and actually using it, through journaling, therapy, or quiet reflection.

K-Hole

Slang for a deep ketamine dive: heavy dissociation, no sense of the body, and the feeling of drifting through abstract mental space.

Looping

A mental effect where the same thought or sensation repeats on a cycle. Common in dissociative and high-dose psychedelic states.

MAO (Monoamine Oxidase)

An enzyme that clears used serotonin, dopamine, and norepinephrine after release. One of the body's built-in brakes on neurotransmitter levels.

MAOI (Monoamine Oxidase Inhibitor)

A substance that blocks MAO, letting serotonin and dopamine build up. Found in ayahuasca and some antidepressants, and prone to dangerous interactions.

Marquis/Mecke/Mandelin

Common reagent test kits that identify substances by the color they turn. Each one flags a different class of drug or adulterant.

Methylation Cycle
A core biochemical loop that repairs DNA and makes neurotransmitters, dependent on B vitamins. Nitrous oxide jams it by knocking out B-12.

Methylcobalamin
The active form of vitamin B-12, and the one nitrous oxide destroys most readily. Preferred for harm reduction, ideally sublingual or injected.

Microdosing
Taking a dose too small to trip on, aiming for a lift in mood, focus, or creativity. Usually LSD or psilocybin.

MMA (Methylmalonic Acid)
A compound that piles up when B-12 is functionally low, even if blood tests look normal. Testing it catches nitrous damage early.

Myelin/Myelination
The fatty insulation around nerve fibers that keeps signals fast and clean. Nitrous erodes it, starting as tingling and risking lasting nerve damage.

NAC (N-Acetylcysteine)
A strong antioxidant and glutathione precursor that lowers oxidative stress. Common in harm reduction for ketamine and stimulant recovery.

Na-R-ALA (Sodium R-Alpha Lipoic Acid)
A potent antioxidant that shields brain cells and supports mitochondria. A staple in protocols countering oxidative stress from MDMA and stimulants.

Neuron

A nerve cell that carries information through electrical and chemical signals, passing messages to other neurons across gaps called synapses.

Neuroplasticity

The brain's ability to rewire itself from experience, learning, or chemistry. It is what makes therapy and psychedelic integration actually stick.

Norepinephrine

Also called noradrenaline. A stress-signal neurotransmitter that lifts heart rate, blood pressure, and alertness. Heavily released by MDMA, and tied to anxiety when chronic.

Oxidative Stress

Cellular wear caused by free radicals, which drug use tends to ramp up. Antioxidants like vitamin C and Na-R-ALA help mop it up.

Preload/Postload

Supplements timed before a session (preload) or after it (postload) to blunt the damage and speed recovery.

Psilocin

The compound psilocybin turns into once you ingest it. Structurally close to serotonin, it hits the same receptors to produce the trip.

Psychedelic

A substance that reshapes perception, mood, and thought by acting on serotonin receptors. The family includes LSD, psilocybin, DMT, and mescaline.

Reagent Test
A chemical spot-test that identifies drugs or adulterants by color. A frontline harm reduction tool for catching fakes and dangerous substitutions.

Reuptake
The neuron reabsorbing the neurotransmitters it just released. This recycling sets how long and how strong a chemical signal lasts.

Reuptake Inhibitor
A substance that blocks reabsorption, so neurotransmitters linger and their effect grows. Cocaine works exactly this way with dopamine.

Roll/Rolling
Slang for being on MDMA: the wave of emotional closeness, easy empathy, and full-body warmth the drug is known for.

Serotonin
A neurotransmitter that manages mood, perception, sleep, appetite, and body temperature. The main lever that psychedelics and SSRIs pull on.

Serotonin Syndrome
A potentially deadly overload of serotonin in the brain. Warning signs include agitation, fever, tremors, and confusion. Treat it as an emergency.

Set and Setting
Your mindset going in (set) and your physical surroundings (setting). Together they shape a psychedelic experience more than the dose does.

SSRI (Selective Serotonin Reuptake Inhibitor)

An antidepressant class, like Zoloft or Lexapro, that raises serotonin by blocking its reuptake. Tends to dull the effects of psychedelics.

Synapse

The tiny gap between two neurons where neurotransmitters cross from one to the next, carrying the signal along.

Synesthesia

A crossing of the senses, where sound takes on color or numbers gain texture. Common on LSD, psilocybin, and DMT.

Thalamus

The brain's sensory relay, filtering incoming signals before they reach awareness. LSD loosens its grip, flooding consciousness with raw, unfiltered input.

THC (Tetrahydrocannabinol)

The main intoxicating compound in cannabis. It produces the high by binding the same CB1 receptors your own endocannabinoids normally use.

Tryptamine

A molecular backbone shared by many psychedelics, including DMT and psilocin, and by serotonin itself. The family they are all built on.

Tyrosine/L-Tyrosine

An amino acid the body turns into dopamine and norepinephrine. Often taken afterward to help restock what a session burned through.

Index

D

E

F

G

H

I

J-K

L

M

N

Q-R

S

T

U-V

W-Z

About the Author

Jeremy Wolfe has spent over a decade teaching people how their brains work, and how not to break them. Known in festival and harm reduction circles as "*Dr. Awkward*,"* he's been giving his pharmacology lectures at Burning Man since 2013 and has volunteered with harm reduction crews at festivals across the country. His background in neuroscience and ethnobotany has made him unusually good at turning dense science into something you can actually use at 2 AM in a field. He speaks at events including PsyCon and lives near Boulder, Colorado. *Altered States* is his first book.

For inquiries, permissions, or speaking engagements, contact: ***jeremy@alteredstatesbook.com*** *or visit* ***alteredstatesbook.com***

* The "Dr." is ceremonial. The "Awkward" was earned.

www.ingramcontent.com/pod-product-compliance
Lightning Source LLC
LaVergne TN
LVHW010717061126
851857LV00014B/1228
9798218936723